Student Solutions

MW00907179

to accompany

Introduction to Linear Algebra with Applications

Jim DeFranza
St. Lawrence University

Daniel Gagliardi
SUNY Canton

McGraw Hill **Higher Education**

Boston   Burr Ridge, IL   Dubuque, IA   New York   San Francisco   St. Louis
Bangkok   Bogotá   Caracas   Kuala Lumpur   Lisbon   London   Madrid   Mexico City
Milan   Montreal   New Delhi   Santiago   Seoul   Singapore   Sydney   Taipei   Toronto

The McGraw·Hill Companies

 **Higher Education**

Student Solutions Manual to accompany
INTRODUCTION TO LINEAR ALGEBRA WITH APPLICATIONS
JIM DEFRANZA AND DANIEL GAGLIARDI

Published by McGraw-Hill Higher Education, an imprint of The McGraw-Hill Companies, Inc., 1221 Avenue of the Americas, New York, NY 10020. Copyright © 2009 by The McGraw-Hill Companies, Inc. All rights reserved.

♲ This book is printed on recycled, acid-free paper containing 10% post consumer waste.

1 2 3 4 5 6 7 8 9 0 QPD/QPD 0 9 8

ISBN: 978-0-07-723959-6
MHID: 0-07-723959-8

www.mhhe.com

Contents

Solutions to All Odd–Numbered Exercises

1 Systems of Linear Equations and Matrices

Exercise Set 1.1

In Section 1.1 of the text, Gaussian Elimination is used to solve a linear system. This procedure utilizes three operations that when applied to a linear system result in a new system that is equivalent to the original. Equivalent means that the linear systems have the same solutions. The three operations are:

- Interchange two equations.

- Multiply any equation by a nonzero constant.

- Add a multiple of one equation to another.

When used judiciously these three operations allow us to reduce a linear system to a triangular linear system, which can be solved. A linear system is consistent if there is at least one solution and is inconsistent if there are no solutions. Every linear system has either a unique solution, infinitely many solutions or no solutions. For example, the triangular linear systems

$$\begin{cases} x_1 - x_2 + x_3 & = 2 \\ x_2 - 2x_3 & = -1 \, , \\ x_3 & = 2 \end{cases} \quad \begin{cases} x_1 - 2x_2 + x_3 & = 2 \\ -x_2 + 2x_3 & = -3 \, , \end{cases} \quad \begin{cases} 2x_1 & + x_3 & = 1 \\ x_2 & - x_3 & = 2 \\ & 0 & = 4 \end{cases}$$

have a unique solution, infinitely many solutions, and no solutions, respectively. In the second linear system, the variable x_3 is a free variable, and once assigned any real number the values of x_1 and x_2 are determined. In this way the linear system has infinitely many solutions. If a linear system has the same form as the second system, but also has the additional equation $0 = 0$, then the linear system will still have free variables. The third system is inconsistent since the last equation $0 = 4$ is impossible. In some cases, the conditions on the right hand side of a linear system are not specified. Consider for example, the linear system

$$\begin{cases} -x_1 - x_2 & = a \\ 2x_1 + 2x_2 + x_3 & = b \ \text{which is equivalent to} \\ 2x_3 & = c \end{cases} \xrightarrow{\hspace{2cm}} \begin{cases} -x_1 - x_2 & = a \\ x_3 & = b + 2a \\ 0 & = c - 2b - 4a \end{cases} .$$

This linear system is consistent only for values a, b and c such that $c - 2b - 4a = 0$.

■ Solutions to Odd Exercises

1. Applying the given operations we obtain the equivalent triangular system

$$\begin{cases} x_1 - x_2 - 2x_3 & = 3 \\ -x_1 + 2x_2 + 3x_3 & = 1 \\ 2x_1 - 2x_2 - 2x_3 & = -2 \end{cases} \xrightarrow{E_1 + E_2 \to E_2} \begin{cases} x_1 - x_2 - 2x_3 & = 3 \\ x_2 + x_3 & = 4 \\ 2x_1 - 2x_2 - 2x_3 & = -2 \end{cases} \xrightarrow{(-2)E_1 + E_3 \to E_3}$$

$$\begin{cases} x_1 - x_2 - 2x_3 & = 3 \\ x_2 + x_3 & = 4 \\ 2x_3 & = -8 \end{cases}$$. Using back substitution, the linear system has the unique solution

$$x_1 = 3, x_2 = 8, x_3 = -4.$$

3. Applying the given operations we obtain the equivalent triangular system

$$\begin{cases} x_1 & + 3x_4 & = 2 \\ x_1 + x_2 & + 4x_4 & = 3 \\ 2x_1 & + x_3 + 8x_4 & = 3 \\ x_1 + x_2 + x_3 & + 6x_4 & = 2 \end{cases} \xrightarrow{(-1)E_1 + E_2 \to E_2} \begin{cases} x_1 & + 3x_4 & = 2 \\ x_2 & + x_4 & = 1 \\ 2x_1 & + x_3 + 8x_4 & = 3 \\ x_1 + x_2 + x_3 & + 6x_4 & = 2 \end{cases}$$

$$\xrightarrow{(-2)E_1 + E_3 \to E_3} \begin{cases} x_1 & + 3x_4 & = 2 \\ x_2 & + x_4 & = 1 \\ +x_3 & + 2x_4 & = -1 \\ x_1 + x_2 + x_3 & + 6x_4 & = 2 \end{cases} \xrightarrow{(-1)E_1 + E_4 \to E_4}$$

$$\begin{cases} x_1 & + 3x_4 & = 2 \\ x_2 & + x_4 & = 1 \\ +x_3 & + 2x_4 & = -1 \\ x_2 + x_3 & + 3x_4 & = 0 \end{cases} \xrightarrow{(-1)E_2 + E_4 \to E_4} \begin{cases} x_1 & + 3x_4 & = 2 \\ x_2 & + x_4 & = 1 \\ x_3 & + 2x_4 & = -1 \\ x_3 & + 2x_4 & = -1 \end{cases}$$

$$\xrightarrow{(-1)E_3 + E_4 \to E_4} \begin{cases} x_1 & + 3x_4 & = 2 \\ x_2 & + x_4 & = 1 \\ x_3 & + 2x_4 & = -1 \\ & 0 & = 0 \end{cases}.$$

The final triangular linear system has more variables than equations, that is, there is a free variable. As a result there are infinitely many solutions. Specifically, using back substitution, the solutions are given by $x_1 = 2 - 3x_4, x_2 = 1 - x_4, x_3 = -1 - 2x_4, x_4 \in \mathbb{R}$.

5. The second equation gives immediately that $x = 0$. Substituting the value $x = 0$ into the first equation, we have that $y = -\frac{2}{3}$. Hence, the linear system has the unique solution $x = 0, y = -\frac{2}{3}$.

7. The first equation gives $x = 1$ and substituting this value in the second equation, we have $y = 0$. Hence, the linear system has the unique solution $x = 1, y = 0$.

9. Notice that the first equation is three times the second and hence, the equations have the same solutions. Since each equation has infinitely many solutions the linear system has infinitely many solutions with solution set $S = \left\{ \left. \left(\frac{2t+4}{3}, t \right) \right| t \in \mathbb{R} \right\}$.

11. The operations $E_1 \leftrightarrow E_3, E_1 + E_2 \to E_2, 3E_1 + E_3 \to E_3$ and $-\frac{8}{5}E_2 + E_3 \to E_3$, reduce the linear system to the equivalent triangular system

$$\begin{cases} x - 2y + z & = -2 \\ -5y + 2z & = -5 \\ \frac{9}{5}z & = 0 \end{cases}.$$

The unique solution is $x = 0, y = 1, z = 0$.

13. The operations $E_1 \leftrightarrow E_2, 2E_1 + E_2 \to E_2, -3E_1 + E_3 \to E_3$ and $E_2 + E_3 \to E_3$, reduce the linear system to the equivalent triangular system

$$\begin{cases} x & + 5z & = -1 \\ -2y + 12z & = -1 \\ 0 & = 0 \end{cases}.$$

The linear system has infinitely many solutions with solution set $S = \left\{ \left(-1 - 5t, 6t + \frac{1}{2}, t\right) \mid t \in \mathbb{R} \right\}$.

15. Adding the two equations yields $6x_1 + 6x_3 = 4$, so that $x_1 = \frac{2}{3} - x_3$. Substituting this value in the first equation gives $x_2 = -\frac{1}{2}$. The linear system has infinitely many solutions with solution set $S = \left\{ \left(-t + \frac{2}{3}, -\frac{1}{2}, t\right) \mid t \in \mathbb{R} \right\}$.

17. The operation $2E_1 + E_2 \to E_2$ gives the equation $-3x_2 - 3x_3 - 4x_4 = -9$. Hence, the linear system has two free variables, x_3 and x_4. The two parameter set of solutions is $S = \left\{ \left(3 - \frac{5}{3}t, -s - \frac{4}{3}t + 3, s, t\right) \mid s, t \in \mathbb{R} \right\}$.

19. The operation $-2E_1 + E_2 \to E_2$ gives $x = b - 2a$. Then $y = a + 2x = a + 2(b - 2a) = 2b - 3a$, so that the unique solution is $x = -2a + b, y = -3a + 2b$.

21. The linear system is equivalent to the triangular linear system

$$\begin{cases} -x & - z & = b \\ & y & = a + 3b \\ & & z & = c - 7b - 2a \end{cases},$$

which has the unique solution $x = 2a + 6b - c, y = a + 3b, z = -2a - 7b + c$.

23. Since the operation $2E_1 + E_2 \to E_2$ gives the equation $0 = 2a + 2$, then the linear system is consistent for $a = -1$.

25. Since the operation $2E_1 + E_2 \to E_2$ gives the equation $0 = a + b$, then the linear system is consistent for $b = -a$.

27. The linear system is equivalent to the triangular linear system

$$\begin{cases} x - 2y + 4z & = a \\ 5y - 9z & = -2a + b \\ 0 & = c - a - b \end{cases}$$

and hence, is consistent for all a, b, and c such that $c - a - b = 0$.

29. The operation $-2E_1 + E_2 \to E_2$ gives the equivalent linear system

$$\begin{cases} x + y & = -2 \\ (a - 2)y & = 7 \end{cases}.$$

Hence, if $a = 2$, the linear system is inconsistent.

31. The operation $-3E_1 + E_2 \to E_2$ gives the equivalent linear system

$$\begin{cases} x - y & = 2 \\ 0 & = a - 6 \end{cases}.$$

Hence, the linear system is inconsistent for all $a \neq 6$.

33. To find the parabola $y = ax^2 + bx + c$ that passes through the specified points we solve the linear system

$$\begin{cases} c & = 0.25 \\ a + b + c & = -1.75 \\ a - b + c & = 4.25 \end{cases}.$$

The unique solution is $a = 1, b = -3$, and $c = \frac{1}{4}$, so the parabola is $y = x^2 - 3x + \frac{1}{4} = \left(x - \frac{3}{2}\right)^2 - 2$. The vertex of the parabola is the point $\left(\frac{3}{2}, -2\right)$.

35. To find the parabola $y = ax^2 + bx + c$ that passes through the specified points we solve the linear system

$$\begin{cases} (-0.5)^2 a - (0.5)b + c & = -3.25 \\ a \quad\quad + \quad b + c & = 2 \\ (2.3)^2 a \quad + (2.3)b + c & = 2.91 \end{cases}.$$

The unique solution is $a = -1, b = -4$, and $c = -1$, so the parabola is $y = -x^2 + 4x - 1 = -(x-2)^2 + 3$. The vertex of the parabola is the point $(2, 3)$.

37. a. The point of intersection of the three lines can be   **b.**
found by solving the linear system

$$\begin{cases} -x + y & = 1 \\ -6x + 5y & = 3 \\ 12x + 5y & = 39 \end{cases}.$$

This linear system has the unique solution $(2, 3)$.

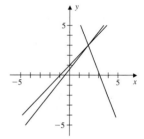

39. a. The linear system $\begin{cases} x + y & = 2 \\ x - y & = 0 \end{cases}$ has the unique solution $x = 1$ and $y = 1$. Notice that the two lines have different slopes.

b. The linear system $\begin{cases} x + y & = 1 \\ 2x + 2y & = 2 \end{cases}$ has infinitely many solutions given by the one parameter set $S = \{(1 - t, t) \mid t \in \mathbb{R}\}$. Notice that the second equation is twice the first and the equations represent the same line.

c. The linear system $\begin{cases} x + y & = 2 \\ 3x + 3y & = -6 \end{cases}$ is inconsistent.

41. a. $S = \{(3 - 2s - t, 2 + s - 2t, s, t) \mid s, t \in \mathbb{R}\}$ **b.** $S = \{(7 - 2s - 5t, s, -2 + s + 2t, t) \mid s, t \in \mathbb{R}\}$

43. Applying $kE_1 \rightarrow E_1, 9E_2 \rightarrow E_2$, and $-E_1 + E_2 \rightarrow E_2$ gives the equivalent linear system

$$\begin{cases} 9kx + \quad\quad k^2 y & = 9k \\ \quad\quad (9 - k^2)y & = -27 - 9k \end{cases}.$$

Whether the linear system is consistent or inconsistent can now be determined by examining the second equation.
a. If $k = 3$, the second equation becomes $0 = -54$, so the linear system is inconsistent. **b.** If $k = -3$, then the second equation becomes $0 = 0$, so the linear system has infinitely many solutions. **c.** If $k \neq \pm 3$, then the linear system has a unique solution.

Exercise Set 1.2

Matrices are used to provide an alternative way to represent a linear system. Reducing a linear system to triangular form is then equivalent to row reducing the augmented matrix corresponding to the linear system to a triangular matrix. For example, the augmented matrix for the linear system

$$\begin{cases} -x_1 - x_2 - x_3 - 2x_4 & = 1 \\ 2x_1 + 2x_2 + x_3 - 2x_4 & = 2 \\ x_1 - 2x_2 + x_3 + 2x_4 & = -2 \end{cases} \text{ is } \begin{bmatrix} -1 & -1 & -1 & -2 & | & 1 \\ 2 & 2 & 1 & -2 & | & 2 \\ 1 & -2 & 1 & 2 & | & -2 \end{bmatrix}.$$

The coefficient matrix is the 3×4 matrix consisting of the coefficients of each variable, that is, the augmented matrix with the augmented column $\begin{bmatrix} 1 \\ 2 \\ -2 \end{bmatrix}$ deleted. The first four columns of the augmented matrix correspond to the variables x_1, x_2, x_4, and x_4, respectively and the augmented column to the constants on the right of each equation. Reducing the linear system using the three valid operations is equivalent to reducing the augmented matrix to a triangular matrix using the row operations:

- Interchange two rows.

- Multiply any row by a nonzero constant.

- Add a multiple of one row to another.

In the above example, the augmented matrix can be reduced to either

$$\left[\begin{array}{cccc|c} \boxed{-1} & -1 & -1 & -2 & 1 \\ 0 & \boxed{-3} & 0 & 0 & -1 \\ 0 & 0 & \boxed{-1} & -6 & 4 \end{array}\right] \quad \text{or} \quad \left[\begin{array}{cccc|c} \boxed{1} & 0 & 0 & -4 & 8/3 \\ 0 & \boxed{1} & 0 & 0 & 1/3 \\ 0 & 0 & \boxed{1} & 6 & -4 \end{array}\right]$$

The left matrix is in row echelon form and the right is in reduced row echelon form. The framed terms are the pivots of the matrix. The pivot entries correspond to dependent variables and the non-pivot entries correspond to free variables. In this example, the free variable is x_4 and x_1, x_2, and x_3 depend on x_4. So the linear system has infinitely many solutions given by $x_1 = \frac{8}{3} + 4x_4, x_2 = \frac{1}{3}, x_3 = -4 - 6x_4$, and x_4 is an arbitrary real number. For a linear system with the same number of equations as variables, there will be a unique solution if and only if the coefficient matrix can be row reduced to the matrix with each diagonal entry 1 and all others 0.

■ Solutions to Odd Exercises

1. $\left[\begin{array}{cc|c} 2 & -3 & 5 \\ -1 & 1 & -3 \end{array}\right]$

3. $\left[\begin{array}{ccc|c} 2 & 0 & -1 & 4 \\ 1 & 4 & 1 & 2 \\ 4 & 1 & -1 & 1 \end{array}\right]$

5. $\left[\begin{array}{ccc|c} 2 & 0 & -1 & 4 \\ 1 & 4 & 1 & 2 \end{array}\right]$

7. $\left[\begin{array}{cccc|c} 2 & 4 & 2 & 2 & -2 \\ 4 & -2 & -3 & -2 & 2 \\ 1 & 3 & 3 & -3 & -4 \end{array}\right]$

9. The linear system has the unique solution $x = -1, y = \frac{1}{2}, z = 0$.

11. The linear system is consistent with free variable z. There are infinitely many solutions given by $x = -3 - 2z, y = 2 + z, z \in \mathbb{R}$.

13. The variable $z = 2$ and y is a free variable, so the linear system has infinitely many solutions given by $x = -3 + 2y, z = 2, y \in \mathbb{R}$

15. The last row of the matrix represents the impossible equation $0 = 1$, so the linear system is inconsistent.

17. The linear system is consistent with free variables z and w. The solutions are given by $x = 3 + 2z - 5w, y = 2 + z - 2w, z \in \mathbb{R}, w \in \mathbb{R}$.

19. The linear system has infinitely many solutions given by $x = 1 + 3w, y = 7 + w, z = -1 - 2w, w \in \mathbb{R}$.

21. The matrix is in reduced row echelon form.

23. Since the matrix contains nonzero entries above the pivots in rows two and three, the matrix is not in reduced row echelon form.

25. The matrix is in reduced row echelon form.

27. Since the first nonzero term in row three is to the left of the first nonzero term in row two, the matrix is not in reduced row echelon form.

29. To find the reduced row echelon form of the matrix we first reduce the matrix to triangular form using $\left[\begin{array}{cc} 2 & 3 \\ -2 & 1 \end{array}\right] \xrightarrow{R_1 + R_2 \to R_2} \left[\begin{array}{cc} 2 & 3 \\ 0 & 4 \end{array}\right]$. The next step is to make the pivots 1, and eliminate the term above the pivot in row two. This gives

$$\begin{bmatrix} 2 & 3 \\ 0 & 4 \end{bmatrix} \xrightarrow{\frac{1}{4}R_2 \to R_2} \begin{bmatrix} 2 & 3 \\ 0 & 1 \end{bmatrix} \xrightarrow{(-3)R_2 + R_1 \to R_1} \begin{bmatrix} 2 & 0 \\ 0 & 1 \end{bmatrix} \xrightarrow{\frac{1}{2}R_1 \to R_1} \begin{bmatrix} 1 & 0 \\ 0 & 1 \end{bmatrix}.$$

31. To avoid the introduction of fractions we interchange rows one and three. The remaining operations are used to change all pivots to ones and eliminate nonzero entries above and below them.

$$\begin{bmatrix} 3 & 3 & 1 \\ 3 & -1 & 0 \\ -1 & -1 & 2 \end{bmatrix} \xrightarrow{R_1 \leftrightarrow R_3} \begin{bmatrix} -1 & -1 & 2 \\ 3 & -1 & 1 \\ 3 & 3 & 1 \end{bmatrix} \xrightarrow{3R_1 + R_2 \to R_2} \begin{bmatrix} -1 & -1 & 2 \\ 0 & -4 & 6 \\ 3 & 3 & 1 \end{bmatrix} \xrightarrow{3R_1 + R_3 \to R_3}$$

$$\begin{bmatrix} -1 & -1 & 2 \\ 0 & -4 & 6 \\ 0 & 0 & 7 \end{bmatrix} \xrightarrow{\frac{1}{7}R_3 \to R_3} \begin{bmatrix} -1 & -1 & 2 \\ 0 & -4 & 6 \\ 0 & 0 & 1 \end{bmatrix} \xrightarrow{(-6)R_3 + R_2 \to R_2} \begin{bmatrix} -1 & -1 & 2 \\ 0 & -4 & 0 \\ 0 & 0 & 1 \end{bmatrix} \xrightarrow{-2R_3 + R_1 \to R_1}$$

$$\begin{bmatrix} -1 & -1 & 0 \\ 0 & -4 & 0 \\ 0 & 0 & 1 \end{bmatrix} \xrightarrow{-\frac{1}{4}R_2 \to R_2} \begin{bmatrix} 1 & -1 & 0 \\ 0 & 1 & 0 \\ 0 & 0 & 1 \end{bmatrix} \xrightarrow{R_2 + R_1 \to R_1} \begin{bmatrix} -1 & 0 & 0 \\ 0 & 1 & 0 \\ 0 & 0 & 1 \end{bmatrix} \xrightarrow{(-1)R_1 \to R_1} \begin{bmatrix} 1 & 0 & 0 \\ 0 & 1 & 0 \\ 0 & 0 & 1 \end{bmatrix}.$$

33. The matrix in reduced row echelon form is $\begin{bmatrix} 1 & 0 & -1 \\ 0 & 1 & 0 \end{bmatrix}.$

35. The matrix in reduced row echelon form is $\begin{bmatrix} 1 & 0 & 0 & -2 \\ 0 & 1 & 0 & -1 \\ 0 & 0 & 1 & 0 \end{bmatrix}.$

37. The augmented matrix for the linear system and the reduced row echelon form are

$$\begin{bmatrix} 1 & 1 & | & 1 \\ 4 & 3 & | & 2 \end{bmatrix} \longrightarrow \begin{bmatrix} 1 & 0 & | & -1 \\ 0 & 1 & | & 2 \end{bmatrix}.$$

The unique solution to the linear system is $x = -1, y = 2.$

39. The augmented matrix for the linear system and the reduced row echelon form are

$$\begin{bmatrix} 3 & -3 & 0 & | & 3 \\ 4 & -1 & -3 & | & 3 \\ -2 & -2 & 0 & | & -2 \end{bmatrix} \longrightarrow \begin{bmatrix} 1 & 0 & 0 & | & 1 \\ 0 & 1 & 0 & | & 0 \\ 0 & 0 & 1 & | & \frac{1}{3} \end{bmatrix}.$$

The unique solution for the linear system is $x = 1, y = 0, z = \frac{1}{3}.$

41. The augmented matrix for the linear system and the reduced row echelon form are

$$\begin{bmatrix} 1 & 2 & 1 & | & 1 \\ 2 & 3 & 2 & | & 0 \\ 1 & 1 & 1 & | & 2 \end{bmatrix} \longrightarrow \begin{bmatrix} 1 & 0 & 1 & | & 0 \\ 0 & 1 & 0 & | & 0 \\ 0 & 0 & 0 & | & 1 \end{bmatrix}.$$

The linear system is inconsistent.

43. The augmented matrix for the linear system and the reduced row echelon form are

$$\begin{bmatrix} 3 & 2 & 3 & | & -3 \\ 1 & 2 & -1 & | & -2 \end{bmatrix} \longrightarrow \begin{bmatrix} 1 & 0 & 2 & | & -\frac{1}{2} \\ 0 & 1 & -\frac{3}{2} & | & -\frac{3}{4} \end{bmatrix}.$$

As a result, the variable x_3 is free and there are infinitely many solutions to the linear system given by $x_1 = -\frac{1}{2} - 2x_3, x_2 = -\frac{3}{4} + \frac{3}{2}x_3, x_3 \in \mathbb{R}.$

45. The augmented matrix for the linear system and the reduced row echelon form are

$$\begin{bmatrix} -1 & 0 & 3 & 1 & | & 2 \\ 2 & 3 & -3 & 1 & | & 2 \\ 2 & -2 & -2 & -1 & | & -2 \end{bmatrix} \longrightarrow \begin{bmatrix} 1 & 0 & 0 & \frac{1}{2} & | & 1 \\ 0 & 1 & 0 & \frac{1}{2} & | & 1 \\ 0 & 0 & 1 & \frac{1}{2} & | & 1 \end{bmatrix}.$$

As a result, the variable x_4 is free and there are infinitely many solutions to the linear system given by $x_1 = 1 - \frac{1}{2}x_4, x_2 = 1 - \frac{1}{2}x_4, x_3 = 1 - \frac{1}{2}x_4, x_4 \in \mathbb{R}$.

47. The augmented matrix for the linear system and the reduced row echelon form are

$$\left[\begin{array}{cccc|c} 3 & -3 & 1 & 3 & -3 \\ 1 & 1 & -1 & -2 & 3 \\ 4 & -2 & 0 & 1 & 0 \end{array}\right] \longrightarrow \left[\begin{array}{cccc|c} 1 & 0 & -\frac{1}{3} & -\frac{1}{2} & 1 \\ 0 & 1 & -\frac{2}{3} & -\frac{3}{2} & 2 \\ 0 & 0 & 0 & 0 & 0 \end{array}\right].$$

As a result, the variables x_3 and x_4 are free and there are infinitely many solutions to the linear system given by $x_1 = 1 + \frac{1}{3}x_3 + \frac{1}{2}x_4, x_2 = 2 + \frac{2}{3}x_3 + \frac{3}{2}x_4, x_3 \in \mathbb{R}, x_4 \in \mathbb{R}$.

49. The augmented matrix for the linear system and the row echelon form are

$$\left[\begin{array}{ccc|c} 1 & 2 & -1 & a \\ 2 & 3 & -2 & b \\ -1 & -1 & 1 & c \end{array}\right] \longrightarrow \left[\begin{array}{ccc|c} 1 & 2 & -1 & a \\ 0 & -1 & 0 & -2a+b \\ 0 & 0 & 0 & -a+b+c \end{array}\right].$$

a. The linear system is consistent precisely when the last equation, from the row echelon form, is consistent. That is, when $c - a + b = 0$. **b.** Similarly, the linear system is inconsistent when $c - a + b \neq 0$. **c.** For those values of a, b, and c for which the linear system is consistent, there is a free variable, so that there are infinitely many solutions. **d.** The linear system is consistent if $a = 1, b = 0, c = 1$. If the variables are denoted by x, y and z, then one solution is obtained by setting $z = 1$, that is, $x = -2, y = 2, z = 1$.

51. The augmented matrix for the linear system and the reduced row echelon form are

$$\left[\begin{array}{ccc|c} -2 & 3 & 1 & a \\ 1 & 1 & -1 & b \\ 0 & 5 & -1 & c \end{array}\right] \longrightarrow \left[\begin{array}{ccc|c} 1 & 0 & -\frac{4}{5} & -\frac{1}{2}a + \frac{3}{10}b \\ 0 & 1 & -\frac{1}{5} & \frac{1}{5}c \\ 0 & 0 & 0 & a + 2b - c \end{array}\right].$$

a. The linear system is consistent precisely when the last equation, from the reduced row echelon form, is consistent. That is, when $a + 2b - c = 0$. **b.** Similarly, the linear system is inconsistent when $a + 2b - c = 0 \neq 0$. **c.** For those values of a, b, and c for which the linear system is consistent, there is a free variable, so that there are infinitely many solutions. **d.** The linear system is consistent if $a = 0, b = 0, c = 0$. If the variables are denoted by x, y and z, then one solution is obtained by setting $z = 1$, that is, $x = \frac{4}{5}, y = \frac{1}{5}, z = 1$.

Exercise Set 1.3

Addition and scalar multiplication are defined componentwise allowing algebra to be performed on expressions involving matrices. Many of the properties enjoyed by the real numbers also hold for matrices. For example, addition is commutative and associative, the matrix of all zeros plays the same role as 0 in the real numbers since the zero matrix added to any matrix A is A. If each component of a matrix A is negated, denoted by $-A$, then $A + (-A)$ is the zero matrix. Matrix multiplication is also defined. The matrix AB that is the product of A with B, is obtained by taking the dot product of each row vector of A with each column vector of B. The order of multiplication is important since it is not always the case that AB and BA are the same matrix. When simplifying expressions with matrices, care is then needed and the multiplication of matrices can be reversed only when it is assumed or known that the matrices commute. The distributive property does hold for matrices, so that $A(B + C) = AB + AC$. In this case however, it is also necessary to note that $(B + C)A = BA + CA$ again since matrix multiplication is not commutative. The transpose of a matrix A, denoted by A^t, is obtained by interchanging the rows and columns of a matrix. There are important properties of the transpose operation you should also be familiar with before solving the exercises. Of particular importance is $(AB)^t = B^t A^t$. Other properties are $(A + B)^t = A^t + B^t, (cA)^t = cA^t$, and $(A^t)^t = A$. A class of matrices that is introduced in Section 1.3 and considered throughout the text are the symmetric matrices. A matrix A is symmetric it is equal to its transpose, that is, $A^t = A$. For example, in the case of 2×2 matrices,

$$A = \left[\begin{array}{cc} a & b \\ c & d \end{array}\right] = A^t = \left[\begin{array}{cc} a & c \\ b & d \end{array}\right] \Leftrightarrow b = c.$$

Here we used that two matrices are equal if and only if corresponding components are equal. Some of the exercises involve showing some matrix or combination of matrices is symmetric. For example, to show that the product of two matrices AB is symmetric, requires showing that $(AB)^t = AB$.

■ Solutions to Odd Exercises

1. Since addition of matrices is defined componentwise, we have that

$$A + B = \begin{bmatrix} 2 & -3 \\ 4 & 1 \end{bmatrix} + \begin{bmatrix} -1 & 3 \\ -2 & 5 \end{bmatrix} = \begin{bmatrix} 2-1 & -3+3 \\ 4-2 & 1+5 \end{bmatrix} = \begin{bmatrix} 1 & 0 \\ 2 & 6 \end{bmatrix}.$$

Also, since addition of real numbers is commutative, $A + B = B + A$.

3. To evaluate the matrix expression $(A + B) + C$ requires we first add $A + B$ and then add C to the result. On the other hand to evaluate $A + (B + C)$ we first evaluate $B + C$ and then add A. Since addition of real numbers is associative the two results are the same, that is $(A + B) + C = \begin{bmatrix} 2 & 1 \\ 7 & 4 \end{bmatrix} = A + (B + C)$.

5. Since a scalar times a matrix multiples each entry of the matrix by the real number, we have that

$$(A - B) + C = \begin{bmatrix} -7 & -3 & 9 \\ 0 & 5 & 6 \\ 1 & -2 & 10 \end{bmatrix} \quad \text{and} \quad 2A + B = \begin{bmatrix} -7 & 3 & 9 \\ -3 & 10 & 6 \\ 2 & 2 & 11 \end{bmatrix}.$$

7. The products are $AB = \begin{bmatrix} 7 & -2 \\ 0 & -8 \end{bmatrix}$ and $BA = \begin{bmatrix} 6 & 2 \\ 7 & -7 \end{bmatrix}$. Notice that, A and B are examples of matrices that do not commute, that is, the order of multiplication can not be reversed.

9. $AB = \begin{bmatrix} -9 & 4 \\ -13 & 7 \end{bmatrix}$ **11.** $AB = \begin{bmatrix} 5 & -6 & 4 \\ 3 & 6 & -18 \\ 5 & -7 & 6 \end{bmatrix}$

13. First, adding the matrices B and C gives

$$\begin{aligned} A(B + C) &= \begin{bmatrix} -2 & -3 \\ 3 & 0 \end{bmatrix} \left(\begin{bmatrix} 2 & 0 \\ -2 & 0 \end{bmatrix} + \begin{bmatrix} 2 & 0 \\ -1 & -1 \end{bmatrix} \right) \\ &= \begin{bmatrix} -2 & -3 \\ 3 & 0 \end{bmatrix} \begin{bmatrix} 4 & 0 \\ -3 & -1 \end{bmatrix} = \begin{bmatrix} (-2)(4) + (-3)(-3) & (-2)(0) + (-3)(-1) \\ (3)(4) + (0)(-3) & (3)(0) + (0)(-1) \end{bmatrix} \\ &= \begin{bmatrix} 1 & 3 \\ 12 & 0 \end{bmatrix}. \end{aligned}$$

15. $2A(B - 3C) = \begin{bmatrix} 10 & -18 \\ -24 & 0 \end{bmatrix}$

17. To find the transpose of a matrix the rows and columns are reversed. So A^t and B^t are 3×2 matrices and the operation is defined. The result is $2A^t - B^t = \begin{bmatrix} 7 & 5 \\ -1 & 3 \\ -3 & -2 \end{bmatrix}$.

19. Since A is 2×3 and B^t is 3×2, then the product AB^t is defined with $AB^t = \begin{bmatrix} -7 & -4 \\ -5 & 1 \end{bmatrix}$

21. $(A^t + B^t)C = \begin{bmatrix} -1 & 7 \\ 6 & 8 \\ 4 & 12 \end{bmatrix}$

23. $(A^t C)B = \begin{bmatrix} 0 & 20 & 15 \\ 0 & 0 & 0 \\ -18 & -22 & -15 \end{bmatrix}$

25. $AB = AC = \begin{bmatrix} -5 & -1 \\ 5 & 1 \end{bmatrix}$

27. The product

$$A^2 = AA = \begin{bmatrix} a & b \\ 0 & c \end{bmatrix} \begin{bmatrix} a & b \\ 0 & c \end{bmatrix} = \begin{bmatrix} a^2 & ab + bc \\ 0 & c^2 \end{bmatrix} = \begin{bmatrix} 1 & 0 \\ 0 & 1 \end{bmatrix}$$

if and only if $a^2 = 1, c^2 = 1$, and $ab + bc = b(a + c) = 0$. That is, $a = \pm 1, c = \pm 1$, and $b(a + c) = 0$, so that A has one of the forms $\begin{bmatrix} 1 & 0 \\ 0 & 1 \end{bmatrix}, \begin{bmatrix} 1 & b \\ 0 & -1 \end{bmatrix}, \begin{bmatrix} -1 & b \\ 0 & 1 \end{bmatrix}$, or $\begin{bmatrix} -1 & 0 \\ 0 & -1 \end{bmatrix}$.

29. Let $A = \begin{bmatrix} 1 & 1 \\ 0 & 0 \end{bmatrix}$ and $B = \begin{bmatrix} -1 & -1 \\ 1 & 1 \end{bmatrix}$. Then $AB = \begin{bmatrix} 0 & 0 \\ 0 & 0 \end{bmatrix}$, and neither of the matrices are the zero matrix. Notice that, this can not happen with real numbers. That is, if the product of two real numbers is zero, then at least one of the numbers must be zero.

31. The product

$$\begin{bmatrix} 1 & 2 \\ a & 0 \end{bmatrix} \begin{bmatrix} 3 & b \\ -4 & 1 \end{bmatrix} = \begin{bmatrix} -5 & b + 2 \\ 3a & ab \end{bmatrix}$$

will equal $\begin{bmatrix} -5 & 6 \\ 12 & 16 \end{bmatrix}$ if and only $b + 2 = 6, 3a = 12$, and $ab = 16$. That is, $a = b = 4$.

33. Several powers of the matrix A are given by

$$A^2 = \begin{bmatrix} 1 & 0 & 0 \\ 0 & 1 & 0 \\ 0 & 0 & 1 \end{bmatrix}, A^3 = \begin{bmatrix} 1 & 0 & 0 \\ 0 & -1 & 0 \\ 0 & 0 & 1 \end{bmatrix}, A^4 = \begin{bmatrix} 1 & 0 & 0 \\ 0 & 1 & 0 \\ 0 & 0 & 1 \end{bmatrix}, \text{ and } A^5 = \begin{bmatrix} 1 & 0 & 0 \\ 0 & -1 & 0 \\ 0 & 0 & 1 \end{bmatrix}.$$

We can see that if n is even, then A^n is the identity matrix, so in particular $A^{20} = \begin{bmatrix} 1 & 0 & 0 \\ 0 & 1 & 0 \\ 0 & 0 & 1 \end{bmatrix}$. Notice also that, if n is odd, then $A^n = \begin{bmatrix} 1 & 0 & 0 \\ 0 & -1 & 0 \\ 0 & 0 & 1 \end{bmatrix}$.

35. We can first rewrite the expression $A^2 B$ as $A^2 B = AAB$. Since $AB = BA$, then $A^2 B = AAB = ABA = BAA = BA^2$.

37. Multiplication of A times the vector $\mathbf{x} = \begin{bmatrix} 1 \\ 0 \\ \vdots \\ 0 \end{bmatrix}$ gives the first column vector of the matrix A. Then $A\mathbf{x} = \mathbf{0}$ forces the first column vector of A to be the zero vector. Then let $\mathbf{x} = \begin{bmatrix} 0 \\ 1 \\ \vdots \\ 0 \end{bmatrix}$ and so on, to show that each column vector of A is the zero vector. Hence, A is the zero matrix.

39. Let $A = \begin{bmatrix} a & b \\ c & d \end{bmatrix}$, so that $A^t = \begin{bmatrix} a & c \\ b & d \end{bmatrix}$. Then

$$AA^t = \begin{bmatrix} a & b \\ c & d \end{bmatrix} \begin{bmatrix} a & c \\ b & d \end{bmatrix} = \begin{bmatrix} a^2 + b^2 & ac + bd \\ ac + bd & c^2 + d^2 \end{bmatrix} = \begin{bmatrix} 0 & 0 \\ 0 & 0 \end{bmatrix}$$

if and only if $a^2 + b^2 = 0, c^2 + d^2 = 0$, and $ac + bd = 0$. The only solution to these equations is $a = b = c = d = 0$, so the only matrix that satisfies $AA^t = \mathbf{0}$ is the 2×2 zero matrix.

41. If A is an $m \times n$ matrix, then A^t is an $n \times m$ matrix, so that AA^t and $A^t A$ are both defined with AA^t being an $m \times m$ matrix and $A^t A$ an $n \times n$ matrix. Since $(AA^t)^t = (A^t)^t A^t = AA^t$, then the matrix AA^t is symmetric. Similarly, $(A^t A)^t = A^t (A^t)^t = A^t A$, so that $A^t A$ is also symmetric.

43. Let $A = (a_{ij})$ be an $n \times n$ matrix. If $A^t = -A$, then the diagonal entries satisfy $a_{ii} = -a_{ii}$ and hence $a_{ii} = 0$ for each i.

Exercise Set 1.4

The inverse of a square matrix plays the same role as the reciprocal of a nonzero number for real numbers. The $n \times n$ identity matrix I, with each diagonal entry a 1 and all other entries 0, satisfies $AI = IA = A$ for all $n \times n$ matrices. Then the inverse of an $n \times n$ matrix, when it exists, is unique and is the matrix, denoted by A^{-1}, such that $AA^{-1} = A^{-1}A = I$. In the case of 2×2 matrices

$$A = \begin{bmatrix} a & b \\ c & d \end{bmatrix} \text{ has an inverse if and only if } ad - bc \neq 0 \text{ and } A^{-1} = \frac{1}{ad - bc} \begin{bmatrix} d & -b \\ -c & a \end{bmatrix}.$$

A procedure for finding the inverse of an $n \times n$ matrix involves forming the augmented matrix $[A \mid I]$ and then row reducing the $n \times 2n$ matrix. If in the reduction process A is transformed to the identity matrix, then the resulting augmented part of the matrix is the inverse. For example, if

$$A = \begin{bmatrix} -2 & -2 & 1 \\ 1 & -1 & -2 \\ 2 & 1 & -2 \end{bmatrix} \text{ and } B = \begin{bmatrix} 0 & -2 & -2 \\ -1 & -1 & 0 \\ 2 & -1 & -1 \end{bmatrix},$$

then A is invertible and B is not since

$$\begin{bmatrix} -2 & -2 & 1 & 1 & 0 & 0 \\ 1 & -1 & -2 & 0 & 1 & 0 \\ 2 & 1 & -2 & 0 & 0 & 1 \end{bmatrix} \rightarrow \begin{bmatrix} 1 & 0 & 0 & -4 & 3 & -5 \\ 0 & 1 & 0 & 2 & -2 & 3 \\ 0 & 0 & 1 & -3 & 2 & -4 \end{bmatrix}$$

$$\underbrace{\qquad\qquad\qquad}_{A^{-1}}$$

but

$$\begin{bmatrix} 0 & -2 & -2 & 1 & 0 & 0 \\ -1 & -1 & 0 & 0 & 1 & 0 \\ 2 & 1 & -1 & 0 & 0 & 1 \end{bmatrix} \rightarrow \begin{bmatrix} 1 & 0 & -1 & 0 & 1 & 1 \\ 0 & 1 & 1 & 0 & -2 & -1 \\ 0 & 0 & 0 & 1 & -4 & -2 \end{bmatrix}.$$

The inverse of the product of two invertible matrices A and B can be found from the inverses of the individual matrices A^{-1} and B^{-1}. But as in the case of the transpose operation, the order of multiplication is reversed, that is, $(AB)^{-1} = B^{-1}A^{-1}$.

■ Solutions to Odd Exercises

1. Since $(1)(-1) - (-2)(3) = 5$ and is nonzero, the inverse exists and $A^{-1} = \frac{1}{5}\begin{bmatrix} -1 & 2 \\ -3 & 1 \end{bmatrix}$.

3. Since $(-2)(-4) - (2)(4) = 0$, then the matrix is not invertible.

5. To determine whether of not the matrix is invertible we row reduce the augmented matrix

$$\begin{bmatrix} 0 & 1 & -1 & 1 & 0 & 0 \\ 3 & 1 & 1 & 0 & 1 & 0 \\ 1 & 2 & -1 & 0 & 0 & 1 \end{bmatrix} \xrightarrow{R_1 \leftrightarrow R_3} \begin{bmatrix} 1 & 2 & -1 & 0 & 0 & 1 \\ 3 & 1 & 1 & 0 & 1 & 0 \\ 0 & 1 & -1 & 1 & 0 & 0 \end{bmatrix} \xrightarrow{(-3)R_1 + R_2 \to R_2}$$

$$\begin{bmatrix} 1 & 2 & -1 & 0 & 0 & 1 \\ 0 & -5 & 4 & 0 & 1 & -3 \\ 0 & 1 & -1 & 1 & 0 & 0 \end{bmatrix} \xrightarrow{R_2 \leftrightarrow R_3} \begin{bmatrix} 1 & 2 & -1 & 0 & 0 & 1 \\ 0 & 1 & -1 & 1 & 0 & 0 \\ 0 & -5 & 4 & 0 & 1 & -3 \end{bmatrix} \xrightarrow{(-2)R_2 + R_1 \to R_1}$$

$$\begin{bmatrix} 1 & 0 & 1 & -2 & 0 & 1 \\ 0 & 1 & -1 & 1 & 0 & 0 \\ 0 & -5 & 4 & 0 & 1 & -3 \end{bmatrix} \xrightarrow{(5)R_2 + R_3 \to R_3} \begin{bmatrix} 1 & 0 & 1 & -2 & 0 & 1 \\ 0 & 1 & -1 & 1 & 0 & 0 \\ 0 & 0 & -1 & 5 & 1 & -3 \end{bmatrix} \xrightarrow{(-1)R_3 \to R_3}$$

$$\begin{bmatrix} 1 & 0 & 1 & -2 & 0 & 1 \\ 0 & 1 & -1 & 1 & 0 & 0 \\ 0 & 0 & 1 & -5 & -1 & 3 \end{bmatrix} \xrightarrow{(-1)R_3 + R_1 \to R_1} \begin{bmatrix} 1 & 0 & 0 & 3 & 1 & -2 \\ 0 & 1 & -1 & 1 & 0 & 0 \\ 0 & 0 & 1 & -5 & -1 & 3 \end{bmatrix} \xrightarrow{(1)R_3 + R_2 \to R_2}$$

$$\left[\begin{array}{ccc|ccc} 1 & 0 & 0 & 3 & 1 & -2 \\ 0 & 1 & 0 & -4 & -1 & 3 \\ 0 & 0 & 1 & -5 & -1 & 3 \end{array}\right].$$ Since the original matrix has been reduce to the identity matrix, the inverse

exists and $A^{-1} = \left[\begin{array}{ccc} 3 & 1 & -2 \\ -4 & -1 & 3 \\ -5 & -1 & 3 \end{array}\right].$

7. Since the matrix A is row equivalent to the matrix $\left[\begin{array}{ccc} 1 & -1 & 0 \\ 0 & 0 & 1 \\ 0 & 0 & 0 \end{array}\right]$, the matrix A can not be reduced to

the identity and hence is not invertible.

9. $A^{-1} = \left[\begin{array}{cccc} 1/3 & -1 & -2 & 1/2 \\ 0 & 1 & 2 & -1 \\ 0 & 0 & -1 & 1/2 \\ 0 & 0 & 0 & -1/2 \end{array}\right]$

11. $A^{-1} = \frac{1}{3}\left[\begin{array}{cccc} 3 & 0 & 0 & 0 \\ -6 & 3 & 0 & 0 \\ 1 & -2 & -1 & 0 \\ 1 & 1 & 1 & 1 \end{array}\right]$

13. The matrix A is not invertible.

15. $A^{-1} = \left[\begin{array}{cccc} 0 & 0 & -1 & 0 \\ 1 & -1 & -2 & 1 \\ 1 & -2 & -1 & 1 \\ 0 & -1 & -1 & 1 \end{array}\right]$

17. Performing the operations, we have that $AB+A = \left[\begin{array}{cc} 3 & 8 \\ 10 & -10 \end{array}\right] = A(B+I)$ and $AB+B = \left[\begin{array}{cc} 2 & 9 \\ 6 & -3 \end{array}\right] = (A+I)B.$

19. Let $A = \left[\begin{array}{cc} 1 & 2 \\ -2 & 1 \end{array}\right]$. **a.** Since $A^2 = \left[\begin{array}{cc} -3 & 4 \\ -4 & -3 \end{array}\right]$ and $-2A = \left[\begin{array}{cc} -2 & -4 \\ 4 & -2 \end{array}\right]$, then $A^2 - 2A + 5I = 0$. **b.**
Since $(1)(1) - (2)(-2) = 5$, the inverse exists and $A^{-1} = \frac{1}{5}\left[\begin{array}{cc} 1 & -2 \\ 2 & 1 \end{array}\right] = \frac{1}{5}(2I - A)$.
c. If $A^2 - 2A + 5I = 0$, then $A^2 - 2A = -5I$, so that $A\left(\frac{1}{5}(2I - A)\right) = \frac{2}{5}A - \frac{1}{5}A^2 = -\frac{1}{5}(A^2 - 2A) = -\frac{1}{5}(-5I) = I$.
Hence $A^{-1} = \frac{1}{5}(2I - A)$.

21. The matrix is row equivalent to $\left[\begin{array}{ccc} 1 & \lambda & 0 \\ 0 & 3-\lambda & 1 \\ 0 & 1-2\lambda & 1 \end{array}\right]$. If $\lambda = -2$, then the second and third rows are
identical, so the matrix can not be row reduced to the identity and hence, is not invertible.

23. a. If $\lambda \neq 1$, then the matrix A is invertible.

b. When $\lambda \neq 1$ the inverse matrix is $A^{-1} = \left[\begin{array}{ccc} -\frac{1}{\lambda-1} & \frac{\lambda}{\lambda-1} & -\frac{\lambda}{\lambda-1} \\ \frac{1}{\lambda-1} & -\frac{1}{\lambda-1} & \frac{1}{\lambda-1} \\ 0 & 0 & 1 \end{array}\right].$

25. The matrices $A = \left[\begin{array}{cc} 1 & 0 \\ 0 & 0 \end{array}\right]$ and $B = \left[\begin{array}{cc} 0 & 0 \\ 0 & 1 \end{array}\right]$ are not invertible, but $A + B = \left[\begin{array}{cc} 1 & 0 \\ 0 & 1 \end{array}\right]$ is invertible.

27. Using the distributive property of matrix multiplication, we have that

$$(A + B)A^{-1}(A - B) = (AA^{-1} + BA^{-1})(A - B) = (I + BA^{-1})(A - B)$$
$$= A - B + B - BA^{-1}B = A - BA^{-1}B.$$

Similarly, $(A - B)A^{-1}(A + B) = A - BA^{-1}B.$

29. a. If A is invertible and $AB = \mathbf{0}$, then $A^{-1}(AB) = A^{-1}\mathbf{0}$, so that $B = \mathbf{0}$.
b. If A is not invertible, then $A\mathbf{x} = \mathbf{0}$ has infinitely many solutions. Let $\mathbf{x_1}, \ldots, \mathbf{x_n}$ be nonzero solutions of
$A\mathbf{x} = \mathbf{0}$ and B the matrix with nth column vector $\mathbf{x_n}$. Then $AB = A\mathbf{x_1} + A\mathbf{x_2} + \cdots + A\mathbf{x_n} = \mathbf{0}$.

31. By the multiplicative property of transpose $(AB)^t = B^t A^t$. Since A and B are symmetric, then $A^t = A$
and $B^t = B$. Hence, $(AB)^t = B^t A^t = BA$. Finally, since $AB = BA$, we have that $(AB)^t = B^t A^t = BA = AB$,
so that AB is symmetric.

33. If $AB = BA$, then $B^{-1}AB = A$, so that $B^{-1}A = AB^{-1}$. Then $(AB^{-1})^t = (B^{-1})^t A^t$. Since $(B^{-1})^t B^t = (BB^{-1})^t = I$, we have that $(B^{-1})^t = (B^t)^{-1}$. Finally, $(AB^{-1})^t = (B^t)^{-1}A^t = B^{-1}A = AB^{-1}$ and hence, AB^{-1} is symmetric.

35. Assuming $A^t = A^{-1}$ and $B^t = B^{-1}$, we need to show that $(AB)^t = (AB)^{-1}$. But $(AB)^t = B^t A^t = B^{-1}A^{-1} = (AB)^{-1}$ and hence, AB is orthogonal.

37. a. Using the associative property of matrix multiplication, we have that

$$(ABC)(C^{-1}B^{-1}A^{-1}) = (AB)CC^{-1}(B^{-1}A^{-1}) = ABB^{-1}A^{-1} = AA^{-1} = I.$$

b. The proof is by induction on the number of matrices k.
Base Case: When $k = 2$, since $(A_1 A_2)^{-1} = A_2^{-1}A_1^{-1}$, the statement holds.
Inductive Hypothesis: Suppose that $(A_1 A_2 \cdots A_k)^{-1} = A_k^{-1}A_{k-1}^{-1} \cdots A_1^{-1}$. Then for $k + 1$ matrices, we have that $(A_1 A_2 \cdots A_k A_{k+1})^{-1} = ([A_1 A_2 \cdots A_k]A_{k+1})^{-1}$. Since $[A_1 A_2 \cdots A_k]$ and A_{k+1} can be considered as two matrices, by the base case, we have that $([A_1 A_2 \cdots A_k]A_{k+1})^{-1} = A_{k+1}^{-1}[A_1 A_2 \cdots A_k]^{-1}$. Finally, by the inductive hypothesis

$$([A_1 A_2 \cdots A_k]A_{k+1})^{-1} = A_{k+1}^{-1}[A_1 A_2 \cdots A_k]^{-1} = A_{k+1}^{-1}A_k^{-1}A_{k-1}^{-1} \cdots A_1^{-1}.$$

39. If A is invertible, then the augmented matrix $[A|I]$ can be row reduced to $[I|A^{-1}]$. If A is upper triangular, then only terms on or above the main diagonal can be affected by the reduction process and hence the inverse is upper triangular. Similarly, the inverse for an invertible lower triangle matrix is also lower triangular.

41. a. Expanding the matrix equation $\begin{bmatrix} a & b \\ c & d \end{bmatrix} \begin{bmatrix} x_1 & x_2 \\ x_3 & x_4 \end{bmatrix} = \begin{bmatrix} 1 & 0 \\ 0 & 1 \end{bmatrix}$, gives $\begin{bmatrix} ax_1 + bx_3 & ax_2 + bx_4 \\ cx_1 + dx_3 & cx_2 + dx_4 \end{bmatrix} = \begin{bmatrix} 1 & 0 \\ 0 & 1 \end{bmatrix}$. **b.** From part (a), we have the two linear systems

$$\begin{cases} ax_1 + bx_3 = 1 \\ cx_1 + dx_3 = 0 \end{cases} \text{ and } \begin{cases} ax_2 + bx_4 = 0 \\ cx_2 + dx_4 = 1 \end{cases}.$$

In the first linear system, multiplying the first equation by d and the second by b and then adding the results gives the equation $(ad - bc)x_1 = d$. Since the assumption is that $ad - bc = 0$, then $d = 0$. Similarly, from the second linear system we conclude that $b = 0$. **c.** From part (b), both $b = 0$ and $d = 0$. Notice that if in addition either $a = 0$ or $c = 0$, then the matrix is not invertible. Also from part (b), we have that $ax_1 = 1, ax_2 = 0, cx_1 = 0$, and $cx_2 = 1$. If a and c are not zero, then these equations are inconsistent and the matrix is not invertible.

Exercise Set 1.5

A linear system can be written as a matrix equation $A\mathbf{x} = \mathbf{b}$, where A is the coefficient matrix of the linear system, $\mathbf{x}$ is the vector of variables and $\mathbf{b}$ is the vector of constants on the right hand side of each equation. For example, the matrix equation corresponding to the linear system

$$\begin{cases} -x_1 - x_2 - x_3 = 2 \\ -x_1 + 2x_2 + 2x_3 = -3 \\ -x_1 + 2x_2 + x_3 = 1 \end{cases} \text{ is } \begin{bmatrix} -1 & -1 & -1 \\ -1 & 2 & 2 \\ -1 & 2 & 1 \end{bmatrix} \begin{bmatrix} x_1 \\ x_2 \\ x_3 \end{bmatrix} = \begin{bmatrix} 2 \\ -3 \\ 1 \end{bmatrix}.$$

If the coefficient matrix A, as in the previous example, has an inverse, then the linear system always has a unique solution. That is, both sides of the equation $A\mathbf{x} = \mathbf{b}$ can be multiplied on the left by A^{-1} to obtain the solution $\mathbf{x} = A^{-1}\mathbf{b}$. In the above example, since the coefficient matrix is invertible the linear system has a unique solution. That is,

$$
\mathbf{x} = \begin{bmatrix} x_1 \\ x_2 \\ x_3 \end{bmatrix} = A^{-1} \begin{bmatrix} 2 \\ -3 \\ 1 \end{bmatrix} = \frac{1}{3} \begin{bmatrix} -2 & -1 & 0 \\ -1 & -2 & 3 \\ 0 & 3 & -3 \end{bmatrix} \begin{bmatrix} 2 \\ -3 \\ 1 \end{bmatrix} = \frac{1}{3} \begin{bmatrix} -1 \\ 7 \\ -12 \end{bmatrix} = \begin{bmatrix} -1/3 \\ 7/3 \\ -4 \end{bmatrix}.
$$

Every homogeneous linear system $A\mathbf{x} = \mathbf{0}$ has at least one solution, namely the trivial solution, where each component of the vector $\mathbf{x}$ is 0. If in addition, the linear system is an $n \times n$ (square) linear system and A is invertible, then the only solution is the trivial one. That is, the unique solution is $\mathbf{x} = A^{-1}\mathbf{0} = \mathbf{0}$. The equivalent statement is that if $A\mathbf{x} = \mathbf{0}$ has two distinct solutions, then the matrix A is not invertible. One additional fact established in the section and that is useful in solving the exercises is that when the linear system $A\mathbf{x} = \mathbf{b}$ has two distinct solutions, then it has infinitely many solutions. That is, every linear linear system has either a unique solution, infinitely many solutions, or is inconsistent.

■ Solutions to Odd Exercises

1. Let $A = \begin{bmatrix} 2 & 3 \\ -1 & 2 \end{bmatrix}$, $\mathbf{x} = \begin{bmatrix} x \\ y \end{bmatrix}$, and $\mathbf{b} = \begin{bmatrix} -1 \\ 4 \end{bmatrix}$.

3. Let $A = \begin{bmatrix} 2 & -3 & 1 \\ -1 & -1 & 2 \\ 3 & -2 & -2 \end{bmatrix}$, $\mathbf{x} = \begin{bmatrix} x \\ y \\ z \end{bmatrix}$, and $\mathbf{b} = \begin{bmatrix} -1 \\ -1 \\ 3 \end{bmatrix}$.

5. Let $A = \begin{bmatrix} 4 & 3 & -2 & -3 \\ -3 & -3 & 1 & 0 \\ 2 & -3 & 4 & -4 \end{bmatrix}$, $\mathbf{x} = \begin{bmatrix} x_1 \\ x_2 \\ x_3 \\ x_4 \end{bmatrix}$, and $\mathbf{b} = \begin{bmatrix} -1 \\ 4 \\ 3 \end{bmatrix}$.

7. $\begin{cases} 2x - 5y = 3 \\ 2x + y = 2 \end{cases}$

9. $\begin{cases} -2y = 3 \\ 2x - y - z = 1 \\ 3x - y + 2z = -1 \end{cases}$

11. $\begin{cases} 2x_1 + 5x_2 - 5x_3 + 3x_4 = 2 \\ 3x_1 + x_2 - 2x_3 - 4x_4 = 0 \end{cases}$

13. The solution is $\mathbf{x} = A^{-1}\mathbf{b} = \begin{bmatrix} 1 \\ 4 \\ -3 \end{bmatrix}$.

15. The solution is

$$
\mathbf{x} = A^{-1}\mathbf{b} = \mathbf{x} = \begin{bmatrix} 9 \\ -3 \\ -8 \\ 7 \end{bmatrix}.
$$

17. The coefficient matrix of the linear system is $A = \begin{bmatrix} 1 & 4 \\ 3 & 2 \end{bmatrix}$, so that

$$
A^{-1} = \frac{1}{(1)(2) - (4)(3)} \begin{bmatrix} 2 & -4 \\ -3 & 1 \end{bmatrix} = -\frac{1}{10} \begin{bmatrix} 2 & -4 \\ -3 & 1 \end{bmatrix}.
$$

Hence, the linear system has the unique solution $\mathbf{x} = \frac{1}{10} \begin{bmatrix} -16 \\ 9 \end{bmatrix}$.

19. If the coefficient matrix is denoted by A, then the unique solution is

$$
A^{-1} \begin{bmatrix} -1 \\ 1 \\ 1 \end{bmatrix} = \begin{bmatrix} 7 & -3 & -1 \\ -3 & 1 & 0 \\ -8 & 3 & 1 \end{bmatrix} \begin{bmatrix} -1 \\ 1 \\ 1 \end{bmatrix} = \begin{bmatrix} -11 \\ 4 \\ 12 \end{bmatrix}.
$$

21. If the coefficient matrix is denoted by A, then the unique solution is

$$A^{-1}\begin{bmatrix} -1 \\ 1 \\ 0 \\ 0 \end{bmatrix} = \begin{bmatrix} 1 & 1 & 0 & 0 \\ -2 & -2 & 1 & -2 \\ 0 & \frac{1}{3} & -\frac{2}{3} & 1 \\ 0 & -\frac{1}{3} & -\frac{1}{3} & 0 \end{bmatrix}\begin{bmatrix} -1 \\ 1 \\ 0 \\ 0 \end{bmatrix} = \frac{1}{3}\begin{bmatrix} 0 \\ 0 \\ 1 \\ -1 \end{bmatrix}.$$

23. a. $\mathbf{x} = \frac{1}{5}\begin{bmatrix} 3 & 1 \\ -2 & 1 \end{bmatrix}\begin{bmatrix} 2 \\ 1 \end{bmatrix} = \frac{1}{5}\begin{bmatrix} 7 \\ -3 \end{bmatrix}$ **b.** $\mathbf{x} = \frac{1}{5}\begin{bmatrix} 3 & 1 \\ -2 & 1 \end{bmatrix}\begin{bmatrix} -3 \\ 2 \end{bmatrix} = \frac{1}{5}\begin{bmatrix} -7 \\ 8 \end{bmatrix}$

25. The reduced row echelon form of the matrix A is

$$\begin{bmatrix} -1 & -4 \\ 3 & 12 \\ 2 & 8 \end{bmatrix} \longrightarrow \begin{bmatrix} 1 & 4 \\ 0 & 0 \\ 0 & 0 \end{bmatrix}$$

hence, the linear system has infinitely many solutions with solution set $S = \left\{ \begin{bmatrix} -4t \\ t \end{bmatrix} \middle| \; t \in \mathbb{R} \right\}$. A particular nontrivial solution is $x = -4$ and $y = 1$.

27. $A = \begin{bmatrix} 1 & 2 & 1 \\ 1 & 2 & 1 \\ 1 & 2 & 1 \end{bmatrix}$
                              29. Since $A\mathbf{u} = A\mathbf{v}$ with $\mathbf{u} \neq \mathbf{v}$, then $A(\mathbf{u} - \mathbf{v}) = \mathbf{0}$. Hence, the homogeneous linear system $A\mathbf{x} = \mathbf{0}$ has a nontrivial solution, so that A is not invertible.

31. a. Let $A = \begin{bmatrix} 2 & 1 \\ -1 & -1 \\ 3 & 2 \end{bmatrix}$, $\mathbf{x} = \begin{bmatrix} x \\ y \end{bmatrix}$, and $\mathbf{b} = \begin{bmatrix} 1 \\ -2 \\ -1 \end{bmatrix}$. The reduced row echelon form of the augmented matrix is

$$\begin{bmatrix} 2 & 1 & | & 1 \\ -1 & 1 & | & -2 \\ 1 & 2 & | & -1 \end{bmatrix} \longrightarrow \begin{bmatrix} 1 & 0 & | & 1 \\ 0 & 1 & | & -1 \\ 0 & 0 & | & 0 \end{bmatrix},$$

so that the solution to the linear system is $\mathbf{x} = \begin{bmatrix} 1 \\ -1 \end{bmatrix}$.   **b.** $C = \frac{1}{3}\begin{bmatrix} 1 & -1 & 0 \\ 1 & 2 & 0 \end{bmatrix}$

c. The solution to the linear system is also given by $C\mathbf{b} = \frac{1}{3}\begin{bmatrix} 1 & -1 & 0 \\ 1 & 2 & 0 \end{bmatrix}\begin{bmatrix} 1 \\ -2 \\ -1 \end{bmatrix} = \begin{bmatrix} 1 \\ -1 \end{bmatrix}$.

Exercise Set 1.6

The determinant of a square matrix is a number that provides information about the matrix. If the matrix is the coefficient matrix of a linear system, then the determinant gives information about the solutions of the linear system. For a 2×2 matrix $A = \begin{bmatrix} a & b \\ c & d \end{bmatrix}$, then $\det(A) = ad - bc$. Another class of matrices where the finding the determinant requires a simple computation are the triangular matrices. In this case the determinant is the product of the entries on the diagonal. So if $A = (a_{ij})$ is an $n \times n$ matrix, then $\det(A) = a_{11}a_{22}\cdots a_{nn}$. The standard row operations on a matrix can be used to reduce a square matrix to an upper triangular matrix and the affect of a row operation on the determinant can be used to find the determinant of the matrix from the triangular form.

- If two rows are interchanged, then the new determinant is the negative of the original determinant.

- If a row is multiplied by a scalar c, then the new determinant is c times the original determinant.

- If a multiple of one row is added to another, then the new determinant is unchanged.

Some immediate consequences of the third property are, if a matrix has a row of zeros, or two equal rows, or one row a multiple of another, then the determinant is 0. The same properties hold if row is replaced with column. It A is an $n \times n$ matrix, since in the matrix cA each row of A is multiplied by c, then $\det(cA) = c^n \det(A)$. Two other useful properties are $\det(A^t) = \det(A)$ and if A is invertible, then $\det(A^{-1}) = \frac{1}{\det(A)}$.

The most important observation made in Section 1.6 is that a square matrix is invertible if and only if its determinant is not zero. Then

$$A \text{ is invertible } \Leftrightarrow \det(A) \neq 0 \Leftrightarrow A\mathbf{x} = \mathbf{b} \text{ has a unique solution}$$
$$\Leftrightarrow A\mathbf{x} = \mathbf{0} \text{ has only the trivial solution}$$
$$\Leftrightarrow A \text{ is row equivalent to } I.$$

One useful observation that follows is that if the determinant of the coefficient matrix is 0, then the linear system is inconsistent or has infinitely many solutions.

■ Solutions to Odd Exercises

1. Since the matrix is triangular, the determinant is the product of the diagonal entries. Hence the determinant is 24.

3. Since the matrix is triangular, the determinant is the product of the diagonal entries. Hence the determinant is -10.

5. Since the determinant is 2, the matrix is invertible.

7. Since the determinant is -6, the matrix is invertible.

9. a. Expanding along row one

$$\det(A) = 2 \begin{vmatrix} -1 & 4 \\ 1 & -2 \end{vmatrix} - (0) \begin{vmatrix} 3 & 4 \\ -4 & -2 \end{vmatrix} + (1) \begin{vmatrix} 3 & -1 \\ -4 & 1 \end{vmatrix} = -5.$$

b. Expanding along row two

$$\det(A) = -3 \begin{vmatrix} 0 & 1 \\ 1 & -2 \end{vmatrix} + (-1) \begin{vmatrix} 2 & 1 \\ -4 & -2 \end{vmatrix} + (4) \begin{vmatrix} 2 & 0 \\ -4 & 1 \end{vmatrix} = -5.$$

c. Expanding along column two

$$\det(A) = -(0) \begin{vmatrix} 3 & 4 \\ -4 & -2 \end{vmatrix} + (-1) \begin{vmatrix} 2 & 1 \\ -4 & -2 \end{vmatrix} + (1) \begin{vmatrix} 2 & 1 \\ 3 & 4 \end{vmatrix} = -5.$$

d. $\det\left(\begin{bmatrix} -4 & 1 & -2 \\ 3 & -1 & 4 \\ 2 & 0 & 1 \end{bmatrix} \right) = 5$

e. Let B denote the matrix in part (d) and B' denote the new matrix. Then $\det(B') = -2\det(B) = -10$. Hence, $\det(A) = \frac{1}{2}\det(B')$.

f. Let B'' denote the new matrix. The row operation does not change the determinant, so $\det(B'') = \det(B') = -10$.

g. Since $\det(A) \neq 0$, the matrix A does have an inverse.

11. Determinant: 13; Invertible

13. Determinant: -16; Invertible

15. Determinant: 0; Not invertible

17. Determinant: 30; Invertible

19. Determinant: -90; Invertible

21. Determinant: 0; Not invertible

23. Determinant: -32; Invertible

25. Determinant: 0; Not invertible

27. Since multiplying a matrix by a scalar multiplies each row by the scalar, we have that $\det(3A) = 3^3 \det(A) = 270$.

29.

$$\det((2A)^{-1}) = \frac{1}{\det(2A)} = \frac{1}{2^3 \det(A)} = \frac{1}{80}$$

31. Expanding along row 3

$$\begin{vmatrix} x^2 & x & 2 \\ 2 & 1 & 1 \\ 0 & 0 & -5 \end{vmatrix} = (-5)\begin{vmatrix} x^2 & x \\ 2 & 1 \end{vmatrix} = -(5)(x^2 - 2x) = -5x^2 + 10x.$$

Then the determinant of the matrix is 0 when $-5x^2 + 10x = -5x(x - 2) = 0$, that is $x = 0$ or $x = 2$.

33. Since the determinant is $a_1 b_2 - b_1 a_2 - xb_2 + xa_2 + yb_1 - ya_1$, then the determinant will be zero precisely when $y = \frac{b_2 - a_2}{b_1 - a_1}x + \frac{b_1 a_2 - a_1 b_2}{b_1 - a_1}$. This equation describes a straight line in the plane.

35. a. $A = \begin{bmatrix} 1 & -1 & -2 \\ -1 & 2 & 3 \\ 2 & -2 & -2 \end{bmatrix}$ **b.** $\det(A) = 2$ **c.** Since the determinant of the coefficient matrix is not zero

it is invertible and hence, the linear system has a unique solution.   **d.** The unique solution is $\mathbf{x} = \begin{bmatrix} 3 \\ 8 \\ -4 \end{bmatrix}$.

37. a. $A = \begin{bmatrix} -1 & 0 & -1 \\ 2 & 0 & 2 \\ 1 & -3 & -3 \end{bmatrix}$ **b.** Expanding along column three, then $\det(A) = \begin{vmatrix} -1 & -1 \\ 2 & 2 \end{vmatrix} = 0.$

c. Since the determinant of the coefficient matrix is 0, A is not invertible. Therefore, the linear system has either no solutions or infinitely many solutions.

d. Since the augmented matrix reduces to

$$\left[\begin{array}{ccc|c} -1 & 0 & -1 & -1 \\ 2 & 0 & 2 & 1 \\ 1 & -3 & -3 & 1 \end{array}\right] \longrightarrow \left[\begin{array}{ccc|c} 1 & 0 & 1 & 0 \\ 0 & 1 & \frac{4}{3} & 0 \\ 0 & 0 & 0 & 1 \end{array}\right]$$

the linear system is inconsistent.

39. a. Since   **b.**

$$\begin{vmatrix} y^2 & x & y & 1 \\ 4 & -2 & -2 & 1 \\ 4 & 3 & 2 & 1 \\ 9 & 4 & -3 & 1 \end{vmatrix} = -29y^2 + 20x - 25y + 106,$$

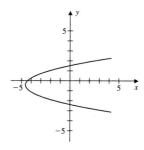

the equation of the parabola is

$$-29y^2 + 20x - 25y + 106 = 0.$$

41. a. Since   **b.**

$$\begin{vmatrix} x^2 & y^2 & x & y & 1 \\ 0 & 16 & 0 & -4 & 1 \\ 0 & 16 & 0 & 4 & 1 \\ 1 & 4 & 1 & -2 & 1 \\ 4 & 9 & 2 & 3 & 1 \end{vmatrix} = 136x^2 - 16y^2 - 328x + 256,$$

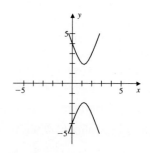

the equation of the hyperbola is

$$136x^2 - 16y^2 - 328x + 256 = 0.$$

43. a. Since

$$\begin{vmatrix} x^2 & xy & y^2 & x & y & 1 \\ 1 & 0 & 0 & -1 & 0 & 1 \\ 0 & 0 & 1 & 0 & 1 & 1 \\ 1 & 0 & 0 & 1 & 0 & 1 \\ 4 & 4 & 4 & 2 & 2 & 1 \\ 9 & 3 & 1 & 3 & 1 & 1 \end{vmatrix} = -12 + 12x^2 - 36xy + 42y^2 - 30y,$$

b.

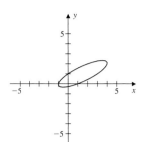

the equation of the ellipse is

$$-12 + 12x^2 - 36xy + 42y^2 - 30y = 0.$$

45.

$$x = \frac{\begin{vmatrix} 7 & -5 \\ 6 & -3 \end{vmatrix}}{\begin{vmatrix} 5 & -5 \\ 2 & -3 \end{vmatrix}} = -\frac{9}{5},$$

$$y = \frac{\begin{vmatrix} 5 & 7 \\ 2 & 6 \end{vmatrix}}{\begin{vmatrix} 5 & -5 \\ 2 & -3 \end{vmatrix}} = -\frac{16}{5}$$

47.

$$x = \frac{\begin{vmatrix} 3 & -4 \\ -10 & 5 \end{vmatrix}}{\begin{vmatrix} -9 & -4 \\ -7 & 5 \end{vmatrix}} = \frac{25}{73},$$

$$y = \frac{\begin{vmatrix} -9 & 3 \\ -7 & -10 \end{vmatrix}}{\begin{vmatrix} -9 & -4 \\ -7 & 5 \end{vmatrix}} = -\frac{111}{73}$$

49. $x = \dfrac{\begin{vmatrix} 4 & -3 \\ 3 & 4 \end{vmatrix}}{\begin{vmatrix} -1 & -3 \\ -8 & 4 \end{vmatrix}} = -\frac{25}{28}, y = \dfrac{\begin{vmatrix} -1 & 4 \\ -8 & 3 \end{vmatrix}}{\begin{vmatrix} -1 & -3 \\ -8 & 4 \end{vmatrix}} = -\frac{29}{28}$

51. $x = \dfrac{\begin{vmatrix} -2 & 3 & 2 \\ -2 & -3 & -8 \\ 2 & 2 & -7 \end{vmatrix}}{\begin{vmatrix} 2 & 3 & 2 \\ -1 & -3 & -8 \\ -3 & 2 & -7 \end{vmatrix}} = -\frac{160}{103}, y = \dfrac{\begin{vmatrix} 2 & -2 & 2 \\ -1 & -2 & -8 \\ -3 & 2 & -7 \end{vmatrix}}{\begin{vmatrix} 2 & 3 & 2 \\ -1 & -3 & -8 \\ -3 & 2 & -7 \end{vmatrix}} = \frac{10}{103}, z = \dfrac{\begin{vmatrix} 2 & 3 & -2 \\ -1 & -3 & -2 \\ -3 & 2 & 2 \end{vmatrix}}{\begin{vmatrix} 2 & 3 & 2 \\ -1 & -3 & -8 \\ -3 & 2 & -7 \end{vmatrix}} = \frac{42}{103}$

53. Expansion of the determinant of A across row one equals the expansion down column one of A^t, so $\det(A) = \det(A^t)$.

Exercise Set 1.7

A factorization of a matrix, like factoring a quadratic polynomial, refers to writing a matrix as the product of other matrices. Just like the resulting linear factors of a quadratic are useful and provide information about the original quadratic polynomial, the lower triangular and upper triangular factors in an LU factorization are easier to work with and can be used to provide information about the matrix. An elementary matrix is obtained by applying one row operation to the identity matrix. For example,

$$\begin{bmatrix} 1 & 0 & 0 \\ 0 & 1 & 0 \\ 0 & 0 & 1 \end{bmatrix} \xrightarrow{(-1)R_1 + R_3 \to R_3} \underbrace{\begin{bmatrix} 1 & 0 & 0 \\ 0 & 1 & 0 \\ -1 & 0 & 1 \end{bmatrix}}_{\text{Elementary Matrix}}.$$

If a matrix A is multiplied by an elementary matrix E, the result is the same as applying to the matrix A the corresponding row operation that defined E. For example, using the elementary matrix above

$$EA = \begin{bmatrix} 1 & 0 & 0 \\ 0 & 1 & 0 \\ -1 & 0 & 1 \end{bmatrix} \begin{bmatrix} 1 & 3 & 1 \\ 2 & 1 & 0 \\ 1 & 2 & 1 \end{bmatrix} = \begin{bmatrix} 1 & 3 & 1 \\ 2 & 1 & 0 \\ 0 & -1 & 0 \end{bmatrix}.$$

Also since each elementary row operation can be reversed, elementary matrices are invertible. To find an LU factorization of A :

- Row reduce the matrix A to an upper triangular matrix U.

- Use the corresponding elementary matrices to write U in the form $U = E_k \cdots E_1 A$.

- If row interchanges are not required, then each of the elementary matrices is lower triangular, so that $A = E_1^{-1} \cdots E_k^{-1} U$ is an LU factorization of A. If row interchanges are required, then a permutation matrix is also required.

When $A = LU$ is an LU factorization of A, and A is invertible, then $A = (LU)^{-1} = U^{-1}L^{-1}$. If the determinant of A is required, then since L and U are triangular, their determinants are simply the product of their diagonal entries and $\det(A) = \det(LU) = \det(L)\det(U)$. An LU factorization can also be used to solve a linear system. To solve the linear system

$$\begin{cases} x_1 - x_2 + 2x_2 & = 2 \\ 2x_1 + 2x_2 + x_3 & = 0 \\ -x_1 + x_2 & = 1 \end{cases}$$

the first step is to find an LU factorization of the coefficient matrix of the linear system. That is,

$$A = LU = \begin{bmatrix} 1 & 0 & 0 \\ 2 & 1 & 0 \\ -1 & 0 & 1 \end{bmatrix} \begin{bmatrix} 1 & -1 & 2 \\ 0 & 4 & -3 \\ 0 & 0 & 2 \end{bmatrix}.$$

Next solve the linear system $L\mathbf{y} = \mathbf{b} = \begin{bmatrix} 2 \\ 0 \\ 1 \end{bmatrix}$ using forward substitution, so that $y_1 = 2, y_2 = -2y_1 = -4, y_3 = 1 + y_1 = 3$. As the final step solve $U\mathbf{x} = \mathbf{y} = \begin{bmatrix} 2 \\ -4 \\ 3 \end{bmatrix}$ using back substitution, so that $x_3 = \frac{3}{2}, x_2 = \frac{1}{4}(-4 + 3x_3) = \frac{1}{8}, x_1 = 2 + x_2 - 2x_3 = -\frac{7}{8}$.

■ Solutions to Odd Exercises

1. a. $E = \begin{bmatrix} 1 & 0 & 0 \\ 2 & 1 & 0 \\ 0 & 0 & 1 \end{bmatrix}$

b. $EA = \begin{bmatrix} 1 & 2 & 1 \\ 5 & 5 & 4 \\ 1 & 1 & -4 \end{bmatrix}$

3. a. $E = \begin{bmatrix} 1 & 0 & 0 \\ 0 & 1 & 0 \\ 0 & -3 & 1 \end{bmatrix}$

b. $EA = \begin{bmatrix} 1 & 2 & 1 \\ 3 & 1 & 2 \\ -8 & -2 & -10 \end{bmatrix}$

5. a. The required row operations are $2R_1 + R_2 \to R_2$, $\frac{1}{10}R_2 \to R_2$, and $-3R_2 + R_1 \to R_1$. The corresponding elementary matrices that transform A to the identity are given in

$$I = E_3 E_2 E_1 A = \begin{bmatrix} 1 & -3 \\ 0 & 1 \end{bmatrix} \begin{bmatrix} 1 & 0 \\ 0 & \frac{1}{10} \end{bmatrix} \begin{bmatrix} 1 & 0 \\ 2 & 1 \end{bmatrix} A.$$

b. Since elementary matrices are invertible, we have that

$$A = E_1^{-1} E_2^{-1} E_3^{-1} = \begin{bmatrix} 1 & 0 \\ -2 & 1 \end{bmatrix} \begin{bmatrix} 1 & 0 \\ 0 & 10 \end{bmatrix} \begin{bmatrix} 1 & 3 \\ 0 & 1 \end{bmatrix}.$$

7. a. The identity matrix can be written as $I = E_5 E_4 E_3 E_2 E_1 A$, where the elementary matrices are

$$E_1 = \begin{bmatrix} 1 & 0 & 0 \\ -2 & 1 & 0 \\ 0 & 0 & 1 \end{bmatrix}, E_2 = \begin{bmatrix} 1 & 0 & 0 \\ 0 & 1 & 0 \\ -1 & 0 & 1 \end{bmatrix}, E_3 = \begin{bmatrix} 1 & -2 & 0 \\ 0 & 1 & 0 \\ 0 & 0 & 1 \end{bmatrix}, E_4 = \begin{bmatrix} 1 & 0 & 11 \\ 0 & 1 & 0 \\ 0 & 0 & 1 \end{bmatrix}, \text{ and}$$

$$E_5 = \begin{bmatrix} 1 & 0 & 0 \\ 0 & 1 & -5 \\ 0 & 0 & 1 \end{bmatrix}.$$ **b.** $A = E_1^{-1} E_2^{-1} E_3^{-1} E_4^{-1} E_5^{-1}$

9. a. The identity matrix can be written as $I = E_6 \cdots E_1 A$, where the elementary matrices are

$$E_1 = \begin{bmatrix} 0 & 1 & 0 \\ 1 & 0 & 0 \\ 0 & 0 & 1 \end{bmatrix}, E_2 = \begin{bmatrix} 1 & -2 & 0 \\ 0 & 1 & 0 \\ 0 & 0 & 1 \end{bmatrix}, E_3 = \begin{bmatrix} 1 & 0 & 0 \\ 0 & 1 & 0 \\ 0 & -1 & 1 \end{bmatrix}, E_4 = \begin{bmatrix} 1 & 0 & 0 \\ 0 & 1 & 1 \\ 0 & 0 & 1 \end{bmatrix},$$

$$E_5 = \begin{bmatrix} 1 & 0 & 1 \\ 0 & 1 & 0 \\ 0 & 0 & 1 \end{bmatrix}, \text{ and } E_6 = \begin{bmatrix} 1 & 0 & 0 \\ 0 & 1 & 0 \\ 0 & 0 & -1 \end{bmatrix}.$$ **b.** $A = E_1^{-1} E_2^{-1} \cdots E_6^{-1}$

11. The matrix A can be row reduced to an upper triangular matrix $U = \begin{bmatrix} 1 & -2 \\ 0 & 1 \end{bmatrix}$, by means of the one operation $3R_1 + R_2 \to R_2$. The corresponding elementary matrix is $E = \begin{bmatrix} 1 & 0 \\ 3 & 1 \end{bmatrix}$, so that $EA = U$. Then the LU factorization of A is $A = LU = E^{-1}U = \begin{bmatrix} 1 & 0 \\ -3 & 1 \end{bmatrix} \begin{bmatrix} 1 & -2 \\ 0 & 1 \end{bmatrix}$.

13. The matrix A can be row reduced to an upper triangular matrix $U = \begin{bmatrix} 1 & 2 & 1 \\ 0 & 1 & 3 \\ 0 & 0 & 1 \end{bmatrix}$, by means of the operations $(-2)R_1 + R_2 \to R_2$ and $3R_1 + R_3 \to R_3$. The corresponding elementary matrices are $E_1 = \begin{bmatrix} 1 & 0 & 0 \\ -2 & 1 & 0 \\ 0 & 0 & 1 \end{bmatrix}$ and $E_2 = \begin{bmatrix} 1 & 0 & 0 \\ 0 & 1 & 0 \\ 3 & 0 & 1 \end{bmatrix}$, so that $E_2 E_1 A = U$. Then the LU factorization of A is $A = LU = E_1^{-1} E_2^{-1} = \begin{bmatrix} 1 & 0 & 0 \\ 2 & 1 & 0 \\ -3 & 0 & 1 \end{bmatrix} \begin{bmatrix} 1 & 2 & 1 \\ 0 & 1 & 3 \\ 0 & 0 & 1 \end{bmatrix}$.

15. $A = LU = \begin{bmatrix} 1 & 0 & 0 \\ 1 & 1 & 0 \\ -1 & -\frac{1}{2} & 1 \end{bmatrix} \begin{bmatrix} 1 & \frac{1}{2} & -3 \\ 0 & 1 & 4 \\ 0 & 0 & 3 \end{bmatrix}$

17. The first step is to determine an LU factorization for the coefficient matrix of the linear system $A = \begin{bmatrix} -2 & 1 \\ 4 & -1 \end{bmatrix}$. We have that $A = LU = \begin{bmatrix} 1 & 0 \\ -2 & 1 \end{bmatrix} \begin{bmatrix} -2 & 1 \\ 0 & 1 \end{bmatrix}$. Next we solve $L\mathbf{y} = \begin{bmatrix} -1 \\ 5 \end{bmatrix}$ to obtain $y_1 = -1$ and $y_2 = 3$. The last step is to solve $U\mathbf{x} = \mathbf{y}$, which has the unique solution $x_1 = 2$ and $x_2 = 3$.

19. An *LU* factorization of the coefficient matrix A is $A = LU = \begin{bmatrix} 1 & 0 & 0 \\ -1 & 1 & 0 \\ 2 & 0 & 1 \end{bmatrix} \begin{bmatrix} 1 & 4 & -3 \\ 0 & 1 & 2 \\ 0 & 0 & 1 \end{bmatrix}$. To solve

$L\mathbf{y} = \begin{bmatrix} 0 \\ -3 \\ 1 \end{bmatrix}$, we have that

$$\left[\begin{array}{ccc|c} 1 & 0 & 0 & 0 \\ -1 & 1 & 0 & -3 \\ 2 & 0 & 1 & 1 \end{array}\right] \longrightarrow \left[\begin{array}{ccc|c} 1 & 0 & 0 & 0 \\ 0 & 1 & 0 & -3 \\ 2 & 0 & 1 & 1 \end{array}\right] \longrightarrow \left[\begin{array}{ccc|c} 1 & 0 & 0 & 0 \\ 0 & 1 & 0 & -3 \\ 0 & 0 & 1 & 1 \end{array}\right]$$

and hence, the solution is $y_1 = 0, y_2 = -3$, and $y_3 = 1$. Finally, the solution to $U\mathbf{x} = \mathbf{y}$, which is the solution to the linear system, is $x_1 = 23, x_2 = -5$, and $x_3 = 1$.

21. *LU* Factorization of the coefficient matrix:

$$A = LU = \begin{bmatrix} 1 & 0 & 0 & 0 \\ 1 & 1 & 0 & 0 \\ 2 & 0 & 1 & 0 \\ -1 & -1 & 0 & 1 \end{bmatrix} \begin{bmatrix} 1 & -2 & 3 & 1 \\ 0 & 1 & 2 & 2 \\ 0 & 0 & 1 & 1 \\ 0 & 0 & 0 & 1 \end{bmatrix}$$

Solution to $L\mathbf{y} = \begin{bmatrix} 5 \\ 6 \\ 14 \\ -8 \end{bmatrix}$: $y_1 = 5, y_2 = 1, y_3 = 4, y_4 = -2$

Solution to $U\mathbf{x} = \mathbf{y}$: $x_1 = -25, x_2 = -7, x_3 = 6, x_4 = -2$

23. In order to row reduce the matrix A to an upper triangular matrix requires the operation of switching rows. This is reflected in the matrix P in the factorization

$$A = PLU = \begin{bmatrix} 0 & 0 & 1 \\ 0 & 1 & 0 \\ 1 & 0 & 0 \end{bmatrix} \begin{bmatrix} 1 & 0 & 0 \\ 2 & 5 & 0 \\ 0 & 1 & -\frac{1}{5} \end{bmatrix} \begin{bmatrix} 1 & -3 & 2 \\ 0 & 1 & -\frac{4}{5} \\ 0 & 0 & 1 \end{bmatrix}.$$

25. Using the *LU* factorization $A = LU = \begin{bmatrix} 1 & 0 \\ -3 & 1 \end{bmatrix} \begin{bmatrix} 1 & 4 \\ 0 & 1 \end{bmatrix}$, we have that

$$A^{-1} = U^{-1}L^{-1} = \begin{bmatrix} 1 & -4 \\ 0 & 1 \end{bmatrix} \begin{bmatrix} 1 & 0 \\ 3 & 1 \end{bmatrix} = \begin{bmatrix} -11 & -4 \\ 3 & 1 \end{bmatrix}.$$

27. Using the *LU* factorization $A = LU = \begin{bmatrix} 1 & 0 & 0 \\ 1 & 1 & 0 \\ 1 & 1 & 1 \end{bmatrix} \begin{bmatrix} 2 & 1 & -1 \\ 0 & 1 & -1 \\ 0 & 0 & 3 \end{bmatrix}$, we have that

$$A^{-1} = U^{-1}L^{-1} = \begin{bmatrix} \frac{1}{2} & -\frac{1}{2} & 0 \\ 0 & 1 & \frac{1}{3} \\ 0 & 0 & \frac{1}{3} \end{bmatrix} \begin{bmatrix} 1 & 0 & 0 \\ -1 & 1 & 0 \\ 0 & -1 & 1 \end{bmatrix} = \begin{bmatrix} 1 & -\frac{1}{2} & 0 \\ -1 & \frac{2}{3} & \frac{1}{3} \\ 0 & -\frac{1}{3} & \frac{1}{3} \end{bmatrix}.$$

29. Suppose

$$\begin{bmatrix} a & 0 \\ b & c \end{bmatrix} \begin{bmatrix} d & e \\ 0 & f \end{bmatrix} = \begin{bmatrix} 0 & 1 \\ 1 & 0 \end{bmatrix}.$$

This gives the system of equations $ad = 0, ae = 1, bd = 1, be + cf = 0$. The first two equations are satisfied only when $a \neq 0$ and $d = 0$. But this incompatible with the third equation.

31. If A is invertible, there are elementary matrices $E_1, \ldots, E_k$ such that $I = E_k \cdots E_1 A$. Similarly, there are elementary matrices $D_1, \ldots, D_\ell$ such that $I = D_\ell \cdots D_1 B$. Then $A = E_k^{-1} \cdots E_1^{-1} D_\ell \cdots D_1 B$, so A is row equivalent to B.

Exercise Set 1.8

1. We need to find positive whole numbers x_1, x_2, x_3, and x_4 such that $x_1 Al_3 + x_2 CuO \longrightarrow x_3 Al_2 O_3 + x_4 Cu$ is balanced. That is, we need to solve the linear system

$$\begin{cases} 3x_1 & = 2x_3 \\ x_2 & = 3x_3 \\ x_2 & = x_4 \end{cases}, \text{ which has infinitely many solutions given by } x_1 = \frac{2}{9}x_2, x_3 = \frac{1}{3}x_2, x_4 = x_2, x_2 \in \mathbb{R}.$$

A particular solution that balances the equation is given by $x_1 = 2, x_2 = 9, x_3 = 3, x_4 = 9$.

3. We need to find positive whole numbers x_1, x_2, x_3, x_4 and x_5 such that

$$x_1 NaHCO_3 + x_2 C_6 H_8 O_7 \longrightarrow x_3 Na_3 C_6 H_5 O_7 + x_4 H_2 O + x_5 CO_2.$$

The augmented matrix for the resulting homogeneous linear system and the reduced row echelon form are

$$\begin{bmatrix} 1 & 0 & -3 & 0 & 0 & | & 0 \\ 1 & 8 & -5 & -2 & 0 & | & 0 \\ 1 & 6 & -6 & 0 & -1 & | & 0 \\ 3 & 7 & -7 & -1 & -2 & | & 0 \end{bmatrix} \longrightarrow \begin{bmatrix} 1 & 0 & 0 & 0 & -1 & | & 0 \\ 0 & 1 & 0 & 0 & -\frac{1}{3} & | & 0 \\ 0 & 0 & 1 & 0 & -\frac{1}{3} & | & 0 \\ 0 & 0 & 0 & 1 & -1 & | & 0 \end{bmatrix}.$$

Hence the solution set for the linear system is given by $x_1 = x_5, x_2 = \frac{1}{3}x_5, x_3 = \frac{1}{3}x_5, x_4 = x_5, x_5 \in \mathbb{R}$. A particular solution that balances the equation is $x_1 = 3, x_2 = 1, x_3 = 1, x_4 = 3, x_5 = 3$.

5. Let $x_1, x_2, \ldots, x_7$ be defined as in the figure. The total flows in and out of the entire network and in and out each intersection are given in the table.

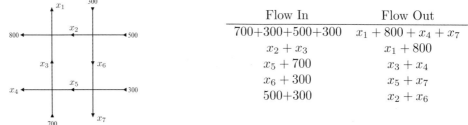

Flow In	Flow Out
700+300+500+300	$x_1 + 800 + x_4 + x_7$
$x_2 + x_3$	$x_1 + 800$
$x_5 + 700$	$x_3 + x_4$
$x_6 + 300$	$x_5 + x_7$
500+300	$x_2 + x_6$

Equating the total flows in and out gives a linear system with solution $x_1 = 1000 - x_4 - x_7, x_2 = 800 - x_6, x_3 = 1000 - x_4 + x_6 - x_7, x_5 = 300 + x_6 - x_7$, with x_4, x_6, and x_7 free variables. Since the network consists of one way streets, the individual flows are nonnegative. As a sample solution let, $x_4 = 200, x_6 = 300, x_7 = 100$, then $x_1 = 700, x_2 = 500, x_3 = 1000, x_5 = 500$.

7. Equating total flows in and out gives the linear system

$$\begin{cases} x_1 & + x_4 & = 150 \\ x_1 - x_2 & - x_5 & = 100 \\ x_2 + x_3 & & = 100 \\ -x_3 & + x_4 + x_5 & = -50 \end{cases}$$

with solution $x_1 = 150 - x_4, x_2 = 50 - x_4 - x_5$, and $x_3 = 50 + x_4 + x_5$. Letting, for example, $x_4 = x_5 = 20$ gives the particular solution $x_1 = 130, x_2 = 10, x_3 = 90$.

9. If x_1, x_2, x_3, and x_4 denote the number of grams required from each of the four food groups, then the specifications yield the linear system

$$\begin{cases} 20x_1 + 30x_2 + 40x_3 + 10x_4 & = 250 \\ 40x_1 + 20x_2 + 35x_3 + 20x_4 & = 300 \\ 50x_1 + 40x_2 + 10x_3 + 30x_4 & = 400 \\ 5x_1 + 5x_2 + 10x_3 + 5x_4 & = 70 \end{cases}.$$

The solution is $x_1 = 1.4, x_2 = 3.2, x_3 = 1.6, x_4 = 6.2$.

11. a. $A = \begin{bmatrix} 0.02 & 0.04 & 0.05 \\ 0.03 & 0.02 & 0.04 \\ 0.03 & 0.3 & 0.1 \end{bmatrix}$ **b.** The internal demand vector is $A \begin{bmatrix} 300 \\ 150 \\ 200 \end{bmatrix} = \begin{bmatrix} 22 \\ 20 \\ 74 \end{bmatrix}$. The total

external demand for the three sectors is $300 - 22 = 278, 150 - 20 = 130$, and $200 - 74 = 126$, respectively.

c. $(I - A)^{-1} \approx \begin{bmatrix} 1.02 & 0.06 & 0.06 \\ 0.03 & 1.04 & 0.05 \\ 0.05 & 0.35 & 1.13 \end{bmatrix}$ **d.** The levels of production that balance the economy are given by

$$X = (I - A)^{-1}D = \begin{bmatrix} 1.02 & 0.06 & 0.06 \\ 0.03 & 1.04 & 0.05 \\ 0.05 & 0.35 & 1.13 \end{bmatrix} \begin{bmatrix} 350 \\ 400 \\ 600 \end{bmatrix} = \begin{bmatrix} 418.2 \\ 454.9 \\ 832.3 \end{bmatrix}.$$

13. a. The health care data are shown in the scatter plot.

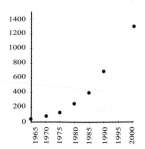

b. If the parabola is $y = ax^2 + bx + c$, then assuming the points $(1970, 80), (1980, 250)$ and $(1990, 690)$ are on

the parabola gives the linear system $\begin{cases} 3880900a + 1970b + c & = 80 \\ 3920400a + 1980b + c & = 250 \\ 3960100a + 1990b + c & = 690 \end{cases}$ **c.** The solution to the linear system

given in part (b) is $a = \frac{27}{20}, b = -\frac{10631}{2}, c = 5232400$, so that the parabola that approximates the data is $y = \frac{27}{20}x^2 - \frac{10631}{2}x + 5232400$.
d.

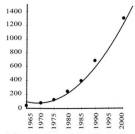

e. The model gives an estimate, in billions of dollars, for health care costs in 2010 at

$$\frac{27}{20}(2010)^2 - \frac{10631}{2}(2010) + 5232400 = 2380.$$

15. a. $A = \begin{bmatrix} 0.9 & 0.08 \\ 0.1 & 0.92 \end{bmatrix}$ **b.** $A \begin{bmatrix} 1500000 \\ 600000 \end{bmatrix} = \begin{bmatrix} 1398000 \\ 702000 \end{bmatrix}$ **c.** $A^2 \begin{bmatrix} 1500000 \\ 600000 \end{bmatrix} = \begin{bmatrix} 1314360 \\ 785640 \end{bmatrix}$ **d.**

$A^n \begin{bmatrix} 1500000 \\ 600000 \end{bmatrix}$

17. The transition matrix is $A = \begin{bmatrix} 0.9 & 0.2 & 0.1 \\ 0.1 & 0.5 & 0.3 \\ 0 & 0.3 & 0.6 \end{bmatrix}$, so the number of the population in each category after

one month are given by $A \begin{bmatrix} 20000 \\ 20000 \\ 10000 \end{bmatrix} = \begin{bmatrix} 23000 \\ 15000 \\ 12000 \end{bmatrix}$, after two months by $A^2 \begin{bmatrix} 20000 \\ 20000 \\ 10000 \end{bmatrix} = \begin{bmatrix} 24900 \\ 13400 \\ 11700 \end{bmatrix}$, and

after one year by $A^{12} \begin{bmatrix} 20000 \\ 20000 \\ 10000 \end{bmatrix} \approx \begin{bmatrix} 30530 \\ 11120 \\ 8350 \end{bmatrix}$.

19. a. $I_1 + I_3 = I_2$ **b.** $\begin{cases} 4I_1 + 3I_2 & = 8 \\ 3I_2 + 5I_3 & = 10 \end{cases}$ **c.** $\begin{cases} I_1 - I_2 & + I_3 & = 0 \\ 4I_1 + 3I_2 & & = 8 \\ & 3I_2 + 5I_3 & = 10 \end{cases}$

The solution to the linear system is $I_1 \approx 0.72, I_2 \approx 1.7, I_3 \approx 0.98$

21. Denote the average temperatures of the four points by a, b, c, and d, clockwise starting with the upper left point. The resulting linear system is

$$\begin{cases} 4a - b - d & = 50 \\ -a + 4b - c & = 55 \\ -b + 4c - d & = 45 \\ -a - c + 4d & = 40. \end{cases}$$

For example, at the first point $a = \frac{20+30+b+d}{4}$. The solution is $a \approx 24.4, b \approx 25.6, c \approx 23.1, d \approx 21.9$.

Review Exercises Chapter 1

1. a. $A = \begin{bmatrix} 1 & 1 & 2 & 1 \\ -1 & 0 & 1 & 2 \\ 2 & 2 & 0 & 1 \\ 1 & 1 & 2 & 3 \end{bmatrix}$ **b.** $\det(A) = -8$ **c.** Since the determinant of the coefficient is not 0,

the matrix is invertible and the linear system is consistent and has a unique solution. **d.** Since the linear system $A\mathbf{x} = \mathbf{b}$ has a unique solution for every $\mathbf{b}$, the only solution to the homogeneous system is the trivial solution. **e.** From part (b), since the determinant is not zero the inverse exists and

$$A^{-1} = \frac{1}{8} \begin{bmatrix} -3 & -8 & -2 & 7 \\ 5 & 8 & 6 & -9 \\ 5 & 0 & -2 & -1 \\ -4 & 0 & 0 & 4 \end{bmatrix}.$$

f. The solution can be found by using the inverse matrix and is given by

$$\mathbf{x} = A^{-1} \begin{bmatrix} 3 \\ 1 \\ -2 \\ 5 \end{bmatrix} = \frac{1}{4} \begin{bmatrix} 11 \\ -17 \\ 7 \\ 4 \end{bmatrix}.$$

3. Let $A = \begin{bmatrix} a & b \\ 0 & c \end{bmatrix}$. The matrix A is idempotent provided $A^2 = A$, that is,

$$\begin{bmatrix} a^2 & ab + bc \\ 0 & c^2 \end{bmatrix} = \begin{bmatrix} a & b \\ 0 & c \end{bmatrix}.$$

The two matrices are equal if and only if the system of equations $a^2 = a, ab + bc = b, c^2 = c$ has a solution. From the first and third equations, we have that $a = 0$ or $a = 1$ and $c = 0$ or $c = 1$. From the second equation, we see that $b(a + c - 1) = 0$, so $b = 0$ or $a + c = 1$. Given these constraints, the possible solutions are $a = 0, c = 0, b = 0; a = 0, c = 1, b \in \mathbb{R}; a = 1, c = 0, b \in \mathbb{R}; a = 1, b = 0, c = 1$

5. a. If $A = \begin{bmatrix} a_1 & b_1 \\ c_1 & d_1 \end{bmatrix}, B = \begin{bmatrix} a_2 & b_2 \\ c_2 & d_2 \end{bmatrix}$, then

$$AB - BA = \begin{bmatrix} a_1a_2 + b_1c_2 & a_1b_2 + b_1d_2 \\ a_2c_1 + c_2d_1 & b_2c_1 + d_1d_2 \end{bmatrix} - \begin{bmatrix} a_1a_2 + b_2c_1 & a_2b_1 + b_2d_1 \\ a_1c_2 + c_1d_2 & b_1c_2 + d_1d_2 \end{bmatrix}$$

$$= \begin{bmatrix} b_1c_2 - b_2c_1 & a_1b_2 + b_1d_1 - a_2b_1 - b_2d_1 \\ a_2c_1 + c_2d_1 - a_1c_2 - c_1d_2 & b_2c_1 - b_1c_2 \end{bmatrix}$$

so that the sum of the diagonal entries is $b_1c_2 - b_2c_1 + b_2c_1 - b_1c_2 = 0$.

b. If M is a 2×2 matrix and the sum of the diagonal entries is 0, then M has the form $M = \begin{bmatrix} a & b \\ c & -a \end{bmatrix}$, then

$$\begin{bmatrix} a & b \\ c & -a \end{bmatrix} \begin{bmatrix} a & b \\ c & -a \end{bmatrix} = \begin{bmatrix} a^2 + bc & 0 \\ 0 & a^2 + bc \end{bmatrix} = (a^2 + bc)I.$$

c. Let $M = AB - BA$. By part (a), $M^2 = kI$, for some k. Then

$$(AB - BA)^2C = M^2C = (kI)C = C(kI) = CM^2 = C(AB - BA)^2.$$

7. a. Since the matrix A is triangular, the determinant is the product of the diagonal entries and hence, $\det(A) = 1$. Since the determinant is not 0, then A is invertible.

b. Six ones can be added making 21 the maximum number of entries that can be one and the matrix is invertible.

9. a. To show that $B = A + A^t$ is symmetric we need to show that $B^t = B$. Using the properties of the transpose operation, we have that $B^t = (A + A^t)^t = A^t + (A^t)^t = A^t + A = B$. Similarly, $C = A - A^t$ is skew-symmetric since $C^t = (A - A^t)^t = A^t - (A^t)^t = A^t - A = -C$.

b. $A = \frac{1}{2}(A + A^t) + \frac{1}{2}(A - A^t)$

Chapter Test Chapter 1

1. T

2. F. All linear systems have either one solution, infinitely many solutions, or no solutions.

3. F. Let $A = \begin{bmatrix} 1 & 1 \\ 1 & 1 \end{bmatrix}$ and $B = \begin{bmatrix} -1 & -1 \\ 1 & 1 \end{bmatrix}$.

4. T

5. F. The linear system $A\mathbf{x} = \mathbf{0}$ has a nontrivial solution if and only if the matrix A is not invertible.

6. T

7. F.

$(ABC)^{-1} = C^{-1}B^{-1}A^{-1}$

8. T

9. T

10. T

11. T

12. T

13. T

14. T

15. F. The determinant is unchanged.

16. T

17. F. Let $A = \begin{bmatrix} 1 & 0 \\ 0 & 1 \end{bmatrix}$, and $B = \begin{bmatrix} -1 & 0 \\ 0 & -1 \end{bmatrix}$

18. T

19. F. The linear system may be inconsistent, but it also may have infinitely many solutions.

20. T

21. F. The inverse is
$$\frac{1}{5} \begin{bmatrix} 1 & 1 \\ -3 & 2 \end{bmatrix}.$$

22. T

23. T

24. T

25. T

26. T

27. F. The determinant of the coefficient matrix is -4.

28. T

29. F. The only solution is
$$x = \frac{5}{4}, y = \frac{1}{4}.$$

30. T

31. T

32. T

33. F. The determinant is given by

$$\begin{vmatrix} 5 & -8 \\ -4 & 6 \end{vmatrix} -2 \begin{vmatrix} 2 & -8 \\ -2 & 6 \end{vmatrix} -3 \begin{vmatrix} 2 & 5 \\ -2 & -4 \end{vmatrix}$$

34. T

35. F

36. T

37. T

38. F.
$$\begin{bmatrix} 1 & 0 & 1 & -1 \\ -1 & -2 & 1 & 3 \\ 0 & 1 & 0 & 3 \end{bmatrix}$$

39. T

40. T

41. T

42. F. The product BA is defined but AB is not defined.

43. T

44. F.
$$\begin{bmatrix} 1 & -8 & 1 \\ -5 & -7 & 16 \end{bmatrix}$$

45. T

2 Linear Combinations and Linear Independence

Exercise Set 2.1

A vector in standard position in the Euclidean spaces $\mathbb{R}^2$ and $\mathbb{R}^3$ is a directed line segment that begins at the origin and ends at some point in either 2 or 3 space. The components of such a vector are the coordinates of the terminal point. In $\mathbb{R}^n$ if vectors are viewed as matrices with n rows and one column, then addition and scalar multiplication of vectors agree with the componentwise operations defined for matrices. In $\mathbb{R}^2$ and $\mathbb{R}^3$ addition and scalar multiplication have the usual geometric meaning as well. For example, $c\mathbf{v}$ changes the length of the vector $\mathbf{v}$ and possibly the direction. If $c > 1$, the length of the vector is increased and the direction is unchanged. If $0 < c < 1$, the length of the vector is decreased and the direction is unchanged. If $c < 0$ the length is changed and the vector is reflected through the origin, so the direction is the opposite of $\mathbf{v}$. Using addition and scalar multiplication of vectors one vector can be obtained from combinations of other vectors. For example, we can ask whether or not the vector

$$\mathbf{v} = \begin{bmatrix} -6 \\ 3 \\ 5 \end{bmatrix}, \text{ can be written in terms of } \begin{bmatrix} -1 \\ 2 \\ 0 \end{bmatrix}, \begin{bmatrix} -1 \\ 2 \\ -1 \end{bmatrix}, \text{ and } \begin{bmatrix} 0 \\ 1 \\ 1 \end{bmatrix}.$$

That is, are there scalars c_1, c_2 and c_3 such that

$$\begin{bmatrix} -6 \\ 3 \\ 5 \end{bmatrix} = c_1 \begin{bmatrix} -1 \\ 2 \\ 0 \end{bmatrix} + c_2 \begin{bmatrix} -1 \\ 2 \\ -1 \end{bmatrix} + c_3 \begin{bmatrix} 0 \\ 1 \\ 1 \end{bmatrix} = \begin{bmatrix} -c_1 - c_2 \\ 2c_1 + 2c_2 + c_3 \\ -c_2 + c_3 \end{bmatrix}?$$

Since two vectors are equal if and only if corresponding components are equal, this leads to the linear system
$$\begin{cases} -c_1 - c_2 = -6 \\ 2c_1 + 2c_2 + c_3 = 3 \\ -c_2 + c_3 = 5 \end{cases}$$. The augmented matrix corresponding to the linear system and the reduced row echelon form are

$$\begin{bmatrix} -1 & -1 & 0 & | & 3 \\ 2 & 2 & 1 & | & -1 \\ 0 & -1 & 1 & | & 2 \end{bmatrix} \longrightarrow \begin{bmatrix} 1 & 0 & 0 & | & -6 \\ 0 & 1 & 0 & | & 3 \\ 0 & 0 & 1 & | & 5 \end{bmatrix},$$

so that the linear system has the unique solution $c_1 = -6, c_2 = 3$ and $c_3 = 5$. As a result the vector can be written as

$$\begin{bmatrix} -6 \\ 3 \\ 5 \end{bmatrix} = -6 \begin{bmatrix} -1 \\ 2 \\ 0 \end{bmatrix} + \begin{bmatrix} -1 \\ 2 \\ -1 \end{bmatrix} + \begin{bmatrix} 0 \\ 1 \\ 1 \end{bmatrix}.$$

Notice that in this example the coefficient matrix of the resulting linear system is always the same regardless of the vector $\mathbf{v}$. As a consequence, every vector $\mathbf{v}$ in $\mathbb{R}^3$ can be written in terms of the three given vectors. If the linear system that is generated turns out to be inconsistent for some vector $\mathbf{v}$, then the vector can not be written as a combination of the others. The linear system can also have infinitely many solutions for a given $\mathbf{v}$, which says the vector can be written in terms of the other vectors in many ways.

■ Solutions to Odd Exercises

1. Adding corresponding components of the vectors $\mathbf{u}$ and $\mathbf{v}$, we have that

$$\mathbf{u} + \mathbf{v} = \begin{bmatrix} -1 \\ 2 \\ 3 \end{bmatrix} = \mathbf{v} + \mathbf{u}.$$

3. $\mathbf{u} - 2\mathbf{v} + 3\mathbf{w} = \begin{bmatrix} 11 \\ -7 \\ 0 \end{bmatrix}$ **5.** $-3(\mathbf{u} + \mathbf{v}) - \mathbf{w} = \begin{bmatrix} 1 \\ -7 \\ -8 \end{bmatrix}$

7. A scalar multiple of a vector is obtained by multiplying each component of the vector by the scalar and addition is componentwise, so that the vector expression simplifies to $\begin{bmatrix} -17 \\ -14 \\ 9 \\ -6 \end{bmatrix}$.

9. To show that $(x_1 + x_2)\mathbf{u} = x_1\mathbf{u} + x_2\mathbf{u}$, we will expand the left hand side to obtain the right hand side. That is,

$$(x_1 + x_2)\mathbf{u} = (x_1 + x_2)\begin{bmatrix} 1 \\ -2 \\ 3 \\ 0 \end{bmatrix} = \begin{bmatrix} x_1 + x_2 \\ -2x_1 - 2x_2 \\ 3x_1 + 3x_2 \\ 0 \end{bmatrix} = \begin{bmatrix} x_1 \\ -2x_1 \\ 3x_1 \\ 0 \end{bmatrix} + \begin{bmatrix} x_2 \\ -2x_2 \\ 3x_2 \\ 0 \end{bmatrix} = x_1\mathbf{u} + x_2\mathbf{v}.$$

11. Simply multiply each of the vectors $\mathbf{e_1}, \mathbf{e_2}$ and $\mathbf{e_3}$ by the corresponding component of the vector. That is, $\mathbf{v} = 2\mathbf{e_1} + 4\mathbf{e_2} + \mathbf{e_3}$

13. $\mathbf{v} = 3\mathbf{e_2} - 2\mathbf{e_3}$

15. Let $\mathbf{w} = \begin{bmatrix} a \\ b \\ c \end{bmatrix}$. Then $-\mathbf{u} + 3\mathbf{v} - 2\mathbf{w} = \mathbf{0}$ if and only if

$$\begin{bmatrix} -1 \\ -4 \\ -2 \end{bmatrix} + \begin{bmatrix} -6 \\ 6 \\ 0 \end{bmatrix} - \begin{bmatrix} 2a \\ 2b \\ 2c \end{bmatrix} = \begin{bmatrix} 0 \\ 0 \\ 0 \end{bmatrix}.$$

Simplifying and equating components gives the equations $-2a - 7 = 0, -2b + 2 = 0$, and $-2c - 2 = 0$ and hence, $\mathbf{w} = \begin{bmatrix} -\frac{7}{2} \\ 1 \\ -1 \end{bmatrix}$.

17. The linear system is $\begin{cases} c_1 & + 3c_2 & = -2 \\ -2c_1 & - 2c_2 & = -1 \end{cases}$, with solution $c_1 = \frac{7}{4}, c_2 = -\frac{5}{4}$. Thus, the vector $\begin{bmatrix} -2 \\ -1 \end{bmatrix}$ can be written as the combination $\frac{7}{4}\begin{bmatrix} 1 \\ -2 \end{bmatrix} - \frac{5}{4}\begin{bmatrix} 3 \\ -2 \end{bmatrix}$.

19. The linear system is $\begin{cases} c_1 - c_2 & = 3 \\ 2c_1 - 2c_2 & = 1 \end{cases}$, which is inconsistent. Hence, the vector $\begin{bmatrix} 3 \\ 1 \end{bmatrix}$ can not be written as a combination of $\begin{bmatrix} 1 \\ 2 \end{bmatrix}$ and $\begin{bmatrix} -1 \\ -2 \end{bmatrix}$.

21. The linear system is $\begin{cases} -4c_1 \quad\quad\; -5c_3 \;\;= -3 \\ 4c_1 + 3c_2 + c_3 \quad\; = -3 \\ 3c_1 - c_2 - 5c_3 \;\;= 4 \end{cases}$. The augmented matrix for the linear system and the

reduced row echelon form are

$$\left[\begin{array}{ccc|c} -4 & 0 & -5 & -3 \\ 4 & 3 & 1 & -3 \\ 3 & -1 & -5 & 4 \end{array}\right] \longrightarrow \left[\begin{array}{ccc|c} 1 & 0 & 0 & \frac{87}{121} \\ 0 & 1 & 0 & -\frac{238}{121} \\ 0 & 0 & 1 & \frac{3}{121} \end{array}\right],$$

so that the unique solution to the linear system is $c_1 = \frac{87}{121}, c_2 = -\frac{238}{121}, c_3 = \frac{3}{121}$. The vector $\begin{bmatrix} -3 \\ -3 \\ 4 \end{bmatrix}$ is a

combination of the three vectors.

23. The linear system is $\begin{cases} -c_1 - c_2 + c_3 \;\;= -1 \\ \quad\quad c_2 - c_3 \;\;= 0 \\ c_1 + c_2 - c_3 \;\;= 2 \end{cases}$, which is inconsistent and hence, the vector $\begin{bmatrix} -1 \\ 0 \\ 2 \end{bmatrix}$ cannot

be written as a combination of the other vectors.

25. All 2×2 vectors. Moreover, $c_1 = \frac{1}{3}a - \frac{2}{3}b, c_2 = \frac{1}{3}a + \frac{1}{3}b$.

27. Row reduction of the matrix $\left[\begin{array}{cc|c} 1 & 2 & a \\ -1 & -2 & b \end{array}\right]$ gives $\left[\begin{array}{cc|c} 1 & 2 & a \\ 0 & 0 & a+b \end{array}\right]$, which is consistent when $b = -a$.

The vector equation can be solved for all vectors of the form $\begin{bmatrix} a \\ -a \end{bmatrix}$ such that $a \in \mathbb{R}$.

29. All 3×3 vectors. Moreover, $c_1 = \frac{1}{3}a - \frac{2}{3}b +$ **31.** All vectors of the form
$\frac{2}{3}c, c_2 = -\frac{1}{3}a + \frac{2}{3}b + \frac{1}{3}c, c_3 = \frac{1}{3}a + \frac{1}{3}b - \frac{1}{3}c$ $\begin{bmatrix} a \\ b \\ 2a-3b \end{bmatrix}$ such that $a, b \in \mathbb{R}$.

33. Let $\mathbf{u} = \begin{bmatrix} u_1 \\ u_2 \\ \vdots \\ u_n \end{bmatrix}, \mathbf{v} = \begin{bmatrix} v_1 \\ v_2 \\ \vdots \\ v_n \end{bmatrix}$, and $\mathbf{w} = \begin{bmatrix} w_1 \\ w_2 \\ \vdots \\ w_n \end{bmatrix}$. Since addition of real numbers is associative, we

have that

$$(\mathbf{u} + \mathbf{v}) + \mathbf{w} = \left(\begin{bmatrix} u_1 + v_1 \\ u_2 + v_2 \\ \vdots \\ u_n + v_n \end{bmatrix}\right) + \begin{bmatrix} w_1 \\ w_2 \\ \vdots \\ w_n \end{bmatrix} = \begin{bmatrix} (v_1 + u_1) + w_1 \\ (v_2 + u_2) + w_2 \\ \vdots \\ (v_n + u_n) + w_n \end{bmatrix} = \begin{bmatrix} u_1 + (v_1 + w_1) \\ u_2 + (v_2 + w_2) \\ \vdots \\ u_n + (v_n + w_n) \end{bmatrix} \mathbf{u} + (\mathbf{v} + \mathbf{w}).$$

35. Let $\mathbf{u} = \begin{bmatrix} u_1 \\ u_2 \\ \vdots \\ u_n \end{bmatrix}$. Then $\mathbf{u} + (-\mathbf{u}) = \begin{bmatrix} u_1 \\ u_2 \\ \vdots \\ u_n \end{bmatrix} + \begin{bmatrix} -u_1 \\ -u_2 \\ \vdots \\ -u_n \end{bmatrix} = \begin{bmatrix} 0 \\ 0 \\ \vdots \\ 0 \end{bmatrix} = (-\mathbf{u}) + \mathbf{u}$. Hence, the vector $-\mathbf{u}$ is

the additive inverse of $\mathbf{u}$.

37. Let $\mathbf{u} = \begin{bmatrix} u_1 \\ u_2 \\ \vdots \\ u_n \end{bmatrix}$ and c and d scalars. Then

$$(c+d)\mathbf{u} = \begin{bmatrix} (c+d)u_1 \\ (c+d)u_2 \\ \vdots \\ (c+d)u_n \end{bmatrix} = \begin{bmatrix} cu_1 + du_1 \\ cu_2 + du_2 \\ \vdots \\ cu_n + du_n \end{bmatrix} = \begin{bmatrix} cu_1 \\ cu_2 \\ \vdots \\ cu_n \end{bmatrix} + \begin{bmatrix} du_1 \\ du_2 \\ \vdots \\ du_n \end{bmatrix} = c\mathbf{u} + d\mathbf{u}.$$

39. Let $\mathbf{u} = \begin{bmatrix} u_1 \\ u_2 \\ \vdots \\ u_n \end{bmatrix}$. Then $(1)\mathbf{u} = \begin{bmatrix} (1)u_1 \\ (1)u_2 \\ \vdots \\ (1)u_n \end{bmatrix} = \mathbf{u}.$

Exercise Set 2.2

The vectors in $\mathbb{R}^2$ and $\mathbb{R}^3$, called the coordinate vectors, are the unit vectors that define the standard axes. These vectors are

$$\mathbf{e_1} = \begin{bmatrix} 1 \\ 0 \end{bmatrix}, \mathbf{e_2} = \begin{bmatrix} 0 \\ 1 \end{bmatrix} \text{ in } \mathbb{R}^2 \text{ and } \mathbf{e_1} = \begin{bmatrix} 1 \\ 0 \\ 0 \end{bmatrix}, \mathbf{e_2} = \begin{bmatrix} 0 \\ 1 \\ 0 \end{bmatrix}, \mathbf{e_3} = \begin{bmatrix} 0 \\ 0 \\ 1 \end{bmatrix} \text{ in } \mathbb{R}^3$$

with the obvious extension to $\mathbb{R}^n$. Every vector in the Euclidean spaces is a combination of the coordinate vectors. For example,

$$\begin{bmatrix} v_1 \\ v_2 \\ v_3 \end{bmatrix} = v_1 \begin{bmatrix} 1 \\ 0 \\ 0 \end{bmatrix} + v_2 \begin{bmatrix} 0 \\ 1 \\ 0 \end{bmatrix} + v_3 \begin{bmatrix} 0 \\ 0 \\ 1 \end{bmatrix}.$$

The coordinate vectors are special vectors but not in the sense of generating all the vectors in the space. Many sets of vectors can generate all other vectors. The vector $\mathbf{v}$ is a linear combination of $\mathbf{v_1}, \mathbf{v_2}, \ldots, \mathbf{v_n}$ if there are scalars $c_1, c_2, \ldots, c_n$ such that

$$\mathbf{v} = c_1\mathbf{v_1} + c_2\mathbf{v_2} + \cdots + c_n\mathbf{v_n}.$$

For specific vectors we have already seen that an equation of this form generates a linear system. If the linear system has at least one solution, then v is a linear combination of the others. Notice that the set of all linear combinations of the vectors $\mathbf{e_1}$ and $\mathbf{e_2}$ in $\mathbb{R}^3$ is the xy plane and hence, not all of $\mathbb{R}^3$. Similarly, all linear combinations of one nonzero vector in $\mathbb{R}^2$ is a line. In the exercises when asked to determine whether or not a vector is a linear combination of other vectors, first set up the equation above and then solve the resulting linear system. For example, to determine all vectors in $\mathbb{R}^3$ that are a linear combination of $\begin{bmatrix} 1 \\ 1 \\ 1 \end{bmatrix}$, $\begin{bmatrix} 0 \\ 1 \\ -1 \end{bmatrix}$,

and $\begin{bmatrix} 2 \\ 5 \\ -1 \end{bmatrix}$ let $\mathbf{v} = \begin{bmatrix} a \\ b \\ c \end{bmatrix}$, be an arbitrary vector. Then form the augmented matrix

$$\begin{bmatrix} 1 & 0 & 2 & | & a \\ 1 & 1 & 5 & | & b \\ 1 & -1 & -1 & | & c \end{bmatrix} \xrightarrow{\text{that row reduces to}} \begin{bmatrix} 1 & 0 & 2 & | & a \\ 0 & 1 & 3 & | & -a+b \\ 0 & 0 & 0 & | & -2a+b-c \end{bmatrix}.$$

Hence, the components of any vector $\mathbf{v}$ that is a linear combination of the three vectors must satisfy $-2a+$

$b - c = 0$. This is not all the vectors in $\mathbb{R}^3$. In this case notice that

$$\begin{bmatrix} 2 \\ 5 \\ -1 \end{bmatrix} = 2 \begin{bmatrix} 1 \\ 1 \\ 1 \end{bmatrix} + 3 \begin{bmatrix} 0 \\ 1 \\ -1 \end{bmatrix}.$$

■ Solutions to Odd Exercises

1. To determine whether or not a vector is a linear combination of other vectors we always set up a vector equation of the form $\mathbf{v} = c_1 + \mathbf{v_1} + \cdots c_n\mathbf{v_n}$ and then determine if the resulting linear system can be solved. In matrix form the linear system is
$\begin{bmatrix} 1 & -2 & \big| & -4 \\ 1 & 3 & \big| & 11 \end{bmatrix} \longrightarrow \begin{bmatrix} 1 & 0 & \big| & 2 \\ 0 & 1 & \big| & 3 \end{bmatrix}$, which has the unique solution $c_1 = 2$ and $c_2 = 3$. Hence, the vector $\mathbf{v}$ is a linear combination of $\mathbf{v_1}$ and $\mathbf{v_2}$.

3. Since the resulting linear system $\begin{bmatrix} -2 & 3 & \big| & 1 \\ 4 & -6 & \big| & 1 \end{bmatrix} \longrightarrow \begin{bmatrix} 1 & -3/2 & \big| & 0 \\ 0 & 0 & \big| & 1 \end{bmatrix}$ is inconsistent, then $\mathbf{v}$ is not a linear combination of $\mathbf{v_1}$ and $\mathbf{v_2}$.

5. Since the resulting linear system

$$\begin{bmatrix} -2 & 1 & \big| & -3 \\ 3 & 4 & \big| & 10 \\ 4 & 2 & \big| & 10 \end{bmatrix} \longrightarrow \begin{bmatrix} 1 & 0 & \big| & 2 \\ 0 & 1 & \big| & 1 \\ 0 & 0 & \big| & 0 \end{bmatrix}$$

is consistent with the unique solution $c_1 = 2$ and $c_2 = 1$, then $\mathbf{v}$ is a linear combination of $\mathbf{v_1}$ and $\mathbf{v_2}$.

7. Since the resulting linear system

$$\begin{bmatrix} 2 & 3 & -2 & \big| & 2 \\ -2 & 0 & 0 & \big| & 8 \\ 0 & -3 & -1 & \big| & 2 \end{bmatrix} \longrightarrow \begin{bmatrix} 1 & 0 & 0 & \big| & -4 \\ 0 & 1 & 0 & \big| & 2/3 \\ 0 & 0 & 1 & \big| & -4 \end{bmatrix},$$

is consistent and has a unique solution, then $\mathbf{v}$ can be written in only one way as a linear combination of $\mathbf{v_1}, \mathbf{v_2}$, and $\mathbf{v_3}$.

9. No.

$$\begin{bmatrix} 1 & -1 & 0 & \big| & -1 \\ 2 & -1 & 1 & \big| & 1 \\ -1 & 3 & 2 & \big| & 5 \end{bmatrix} \longrightarrow \begin{bmatrix} 1 & 0 & 0 & \big| & 0 \\ 0 & 1 & 1 & \big| & 0 \\ 0 & 0 & 0 & \big| & 1 \end{bmatrix}$$

11. Yes, there is a unique solution.

$$\begin{bmatrix} 2 & 1 & -1 & \big| & 3 \\ -3 & 6 & -1 & \big| & -17 \\ 4 & -1 & 2 & \big| & 17 \\ 1 & 2 & 3 & \big| & 7 \end{bmatrix} \longrightarrow \begin{bmatrix} 1 & 0 & 0 & \big| & 3 \\ 0 & 1 & 0 & \big| & -1 \\ 0 & 0 & 1 & \big| & 2 \\ 0 & 0 & 0 & \big| & 0 \end{bmatrix}$$

13. Infinitely many ways with scalars given by $c_1 = 1 + \frac{1}{3}c_3, c_2 = 1 + \frac{7}{3}c_3, c_3 \in \mathbb{R}$.

15. Infinitely many ways with scalars given by $c_1 = 3 + 6c_4, c_2 = -2 - c_4, c_3 = 2 + 2c_4, c_4 \in \mathbb{R}$.

17. The matrix equation

$$c_1 M_1 + c_2 M_2 + c_3 M_3 = c_1 \begin{bmatrix} 1 & 2 \\ 1 & -1 \end{bmatrix} + c_2 \begin{bmatrix} -2 & 3 \\ 1 & 4 \end{bmatrix} + c_3 \begin{bmatrix} -1 & 3 \\ 2 & 1 \end{bmatrix} = \begin{bmatrix} -2 & 4 \\ 4 & 0 \end{bmatrix}$$

leads to the augmented matrix

$$\begin{bmatrix} 1 & -2 & -1 & \big| & -2 \\ 2 & 3 & 3 & \big| & 4 \\ 1 & 1 & 2 & \big| & 4 \\ -1 & 4 & 1 & \big| & 0 \end{bmatrix} \longrightarrow \begin{bmatrix} 1 & 0 & 0 & \big| & -1 \\ 0 & 1 & 0 & \big| & -1 \\ 0 & 0 & 1 & \big| & 3 \\ 0 & 0 & 0 & \big| & 0 \end{bmatrix},$$

and hence, the linear system has the unique solution $c_1 = -1, c_2 = -1$, and $c_3 = 3$. Consequently, the matrix M is a linear combination of the three matrices.

19. The matrix M is not a linear combination of the three matrices since

$$\left[\begin{array}{ccc|c} 2 & 3 & 3 & 2 \\ 2 & -1 & -1 & 1 \\ -1 & 2 & 2 & -1 \\ 3 & -2 & 2 & 2 \end{array}\right] \longrightarrow \left[\begin{array}{ccc|c} 1 & 0 & 0 & 0 \\ 0 & 1 & 0 & 0 \\ 0 & 0 & 1 & 0 \\ 0 & 0 & 0 & 1 \end{array}\right],$$

so that the linear system is inconsistent.

21. $A\mathbf{x} = 2\left[\begin{array}{c} 1 \\ -2 \end{array}\right] - \left[\begin{array}{c} 3 \\ 1 \end{array}\right]$

23.

$$(\mathbf{AB})_1 = 3\left[\begin{array}{c} -1 \\ 3 \end{array}\right] + 2\left[\begin{array}{c} -2 \\ 4 \end{array}\right]$$

$$(\mathbf{AB})_2 = 2\left[\begin{array}{c} -1 \\ 3 \end{array}\right] + 5\left[\begin{array}{c} -2 \\ 4 \end{array}\right]$$

25. The linear combination $c_1(1+x) + c_2(x^2) = 2x^2 - 3x - 1$ if and only if $c_1 + c_1 x + c_2 x^2 = 2x^2 - 3x - 1$. These two polynomials will agree for all x if and only if the coefficients of like terms agree. That is, if and only if $c_1 = -1, c_1 = -3$, and $c_2 = 2$, which is not possible. Therefore, the polynomial $p(x)$ can not be written as a linear combination of $1 + x$ and x^2.

27. Consider the equation $c_1(1+x) + c_2(-x) + c_3(x^2+1) + c_4(2x^3 - x + 1) = x^3 - 2x + 1$, which is equivalent to

$$(c_1 + c_3 + c_4) + (c_1 - c_2 - c_4)x + c_3 x^2 + 2c_4 x^3 = x^3 - 2x + 1.$$

After equating the coefficients of like terms, we have the linear system $\begin{cases} c_1 + c_3 + c_4 = 1 \\ c_1 - c_2 - c_4 = -2 \\ c_3 = 0 \\ 2c_4 = 1 \end{cases}$, which has

the unique solution $c_1 = \frac{1}{2}, c_2 = 2, c_3 = 0$, and $c_4 = \frac{1}{2}$. Hence, $x^3 - 2x + 1 = \frac{1}{2}(1+x) + 2(-x) + 0(x^2+1) + \frac{1}{2}(2x^3 - x + 1)$

29. Since

$$\left[\begin{array}{ccc|c} 1 & 3 & 1 & a \\ 2 & 7 & 2 & b \\ -1 & -2 & 0 & c \end{array}\right] \longrightarrow \left[\begin{array}{ccc|c} 1 & 3 & 1 & a \\ 0 & 1 & 1 & -2a+b \\ 0 & 0 & 0 & 3a - b + c \end{array}\right],$$

all vectors $\left[\begin{array}{c} a \\ b \\ c \end{array}\right]$ such that $3a - b + c = 0$ can be written as a linear combination of the three given vectors.

31.

$$\mathbf{v} = \mathbf{v_1} + \mathbf{v_2} + \mathbf{v_3} + \mathbf{v_4}$$
$$= \mathbf{v_1} + \mathbf{v_2} + \mathbf{v_3} + (\mathbf{v_1} - 2\mathbf{v_2} + 3\mathbf{v_3})$$
$$= 2\mathbf{v_1} - \mathbf{v_2} + 4\mathbf{v_3}$$

33. Since $c_1 \neq 0$, then

$$\mathbf{v_1} = -\frac{c_2}{c_1}\mathbf{v_2} - \cdots - \frac{c_n}{c_1}\mathbf{v_n}.$$

35. In order to show that $S_1 = S_2$, we will show that each is a subset of the other. Let $\mathbf{v} \in S_1$, so that there are scalars $c_1, \ldots, c_k$ such that $\mathbf{v} = c_1\mathbf{v_1} + \cdots + c_k\mathbf{v_k}$. Since $c \neq 0$, then $\mathbf{v} = c_1\mathbf{v_1} + \cdots + \frac{c_k}{c}(c\mathbf{v_k})$, so $\mathbf{v} \in S_2$. If $\mathbf{v} \in S_2$, then $\mathbf{v} = c_1\mathbf{v_1} + \cdots + (cc_k)\mathbf{v_k}$, so $\mathbf{v} \in S_1$. Therefore $S_1 = S_2$.

37. If $\mathbf{A}_3 = c\mathbf{A}_1$, then $\det(A) = 0$. Since the linear system is assumed to be consistent, then it must have infinitely many solutions.

39. If $f(x) = e^x$ and $g(x) = e^{\frac{1}{2}x}$, then $f'(x) = e^x = f''(x)$, $g'(x) = \frac{1}{2}e^{\frac{1}{2}x}$, and $g''(x) = \frac{1}{4}e^{\frac{1}{2}x}$. Then $2f'' - 3f' + f = 2e^x - 3e^x + e^x = 0$ and $2g'' - 3g' + g = \frac{1}{2}e^{\frac{1}{2}x} - \frac{3}{2}e^{\frac{1}{2}x} + e^{\frac{1}{2}x} = 0$, and hence $f(x)$ and $g(x)$ are solutions to the differential equation. In a similar manner, for arbitrary constants c_1 and c_2, the function $c_1 f(x) + c_2 g(x)$ is also a solution to the differential equation.

Exercise Set 2.3

In Section 2.3, the fundamental concept of linear independence is introduced. In $\mathbb{R}^2$ and $\mathbb{R}^3$, two nonzero vectors are linearly independent if and only if they are not scalar multiples of each other, so they do not lie on the same line. To determine whether or not a set of vectors $S = \{\mathbf{v_1}, \mathbf{v_2}, \dots, \mathbf{v_k}\}$ is linearly independent set up the vector equation

$$c_1 \mathbf{v_1} + c_2 \mathbf{v_2} + \cdots + c_k \mathbf{v_k} = \mathbf{0}.$$

If the only solution to the resulting system of equations is $c_1 = c_2 = \cdots = c_k = 0$, then the vectors are linearly independent. If there is one or more nontrivial solutions, then the vectors are linearly dependent. For example, the coordinate vectors in Euclidean space are linearly independent. An alternative method, for determining linear independence is to form a matrix A with column vectors the vectors to test. The matrix must be a square matrix, so for example, if the vectors are in $\mathbb{R}^4$, then there must be four vectors.

- If $\det(A) \neq 0$, then the vectors are linearly independent.

- If $\det(A) = 0$, then the vectors are linearly dependent.

For example, to determine whether or not the vectors $\begin{bmatrix} -1 \\ 1 \\ 3 \end{bmatrix}$, $\begin{bmatrix} 2 \\ 1 \\ 2 \end{bmatrix}$, and $\begin{bmatrix} 3 \\ 5 \\ -1 \end{bmatrix}$ are linearly independent start with the equation

$$c_1 \begin{bmatrix} -1 \\ 1 \\ 3 \end{bmatrix} + c_2 \begin{bmatrix} 2 \\ 1 \\ 2 \end{bmatrix} + c_3 \begin{bmatrix} 3 \\ 5 \\ -1 \end{bmatrix} = \begin{bmatrix} 0 \\ 0 \\ 0 \end{bmatrix}.$$

This yields the linear system

$$\begin{cases} -c_1 + 2c_2 + 3c_3 = 0 \\ c_1 + c_2 + 5c_3 = 0 \\ 3c_1 + 2c_2 - c_3 = 0 \end{cases} \text{ with augmented matrix } \begin{bmatrix} -1 & 2 & 3 & | & 0 \\ 1 & 1 & 5 & | & 0 \\ 3 & 2 & -1 & | & 0 \end{bmatrix} \xrightarrow{\text{which reduces to}} \begin{bmatrix} -1 & 2 & 3 & | & 0 \\ 0 & 1 & 1 & | & 0 \\ 0 & 0 & 5 & | & 0 \end{bmatrix}.$$

So the only solution is $c_1 = c_2 = c_3 = 0$, and the vectors are linearly independent. Now notice that the coefficient matrix

$$A = \begin{bmatrix} -1 & 2 & 3 \\ 1 & 1 & 3 \\ 3 & 2 & -1 \end{bmatrix} \xrightarrow{\text{reduces further to}} \begin{bmatrix} 1 & 0 & 0 \\ 0 & 1 & 0 \\ 0 & 0 & 1 \end{bmatrix},$$

so that A is row equivalent to the identity matrix. This implies the inverse A^{-1} exists and that $\det(A) \neq 0$. So we could have computed $\det(A)$ to conclude the vectors are linearly independent. In addition, since A^{-1} exists the linear system $A \begin{bmatrix} c_1 \\ c_2 \\ c_3 \end{bmatrix} = \mathbf{b}$ has a unique solution for every vector $\mathbf{b}$ in $\mathbb{R}^3$. So ever vector in $\mathbb{R}^3$ can be written uniquely as a linear combination of the three vectors. The uniqueness is a key result of the linear independence. Other results that aid in making a determination of linear independence or dependence of a set S are:

- If the zero vector is in S, then S is linearly dependent.

- If S consists of m vectors in $\mathbb{R}^n$ and $m > n$, then S is linearly dependent.

- At least one vector in S is a linear combination of other vectors in S if and only if S is linearly dependent.

- Any subset of a set of linearly independent vectors is linearly independent.

- If S a linearly dependent set and is contained by another set T, then T is also linearly dependent.

■ Solutions to Odd Exercises

1. Since $\begin{vmatrix} -1 & 2 \\ 1 & -3 \end{vmatrix} = 1$, the vectors are linearly independent.

3. Since $\begin{vmatrix} 1 & -2 \\ -4 & 8 \end{vmatrix} = 0$, the vectors are linearly dependent.

5. To solve the linear system $c_1 \begin{bmatrix} -1 \\ 2 \\ 1 \end{bmatrix} + c_2 \begin{bmatrix} 2 \\ 2 \\ 3 \end{bmatrix} = \begin{bmatrix} 0 \\ 0 \\ 0 \end{bmatrix}$, we have that

$\begin{bmatrix} -1 & 2 \\ 2 & 2 \\ 1 & 3 \end{bmatrix} \longrightarrow \begin{bmatrix} -1 & 2 \\ 0 & 6 \\ 0 & 0 \end{bmatrix}$, so the only solution is the trivial solution and hence, the vectors are linearly independent.

7. Since $\begin{vmatrix} -4 & -5 & 3 \\ 4 & 3 & -5 \\ -1 & 3 & 5 \end{vmatrix} = 0$, the vectors are linearly dependent.

9. Since $\begin{bmatrix} 3 & 1 & 3 \\ -1 & 0 & -1 \\ -1 & 2 & 0 \\ 2 & 1 & 1 \end{bmatrix} \longrightarrow \begin{bmatrix} 3 & 1 & 3 \\ 0 & 1/3 & 0 \\ 0 & 0 & 1 \\ 0 & 0 & 0 \end{bmatrix}$, the linear system $c_1 \mathbf{v_1} + c_2 \mathbf{v_2} + c_3 \mathbf{v_3} = \mathbf{0}$ has only the trivial solution and hence, the vectors are linearly independent.

11. From the linear system $c_1 \begin{bmatrix} 3 & 3 \\ 2 & 1 \end{bmatrix} + c_2 \begin{bmatrix} 0 & 1 \\ 0 & 0 \end{bmatrix} + c_3 \begin{bmatrix} 1 & -1 \\ -1 & -2 \end{bmatrix} = \begin{bmatrix} 0 & 0 \\ 0 & 0 \end{bmatrix}$, we have that

$\begin{bmatrix} 3 & 0 & 1 \\ 3 & 1 & -1 \\ 2 & 0 & -1 \\ 1 & 0 & -2 \end{bmatrix} \longrightarrow \begin{bmatrix} 3 & 0 & 1 \\ 0 & 1 & -2 \\ 0 & 0 & -5/3 \\ 0 & 0 & 0 \end{bmatrix}$. Since the homogeneous linear system has only the trivial solution, then the matrices are linearly independent.

13. Since $\begin{bmatrix} 1 & 0 & -1 & 1 \\ -2 & -1 & 1 & 1 \\ -2 & 2 & -2 & -1 \\ -2 & 2 & 2 & -2 \end{bmatrix} \longrightarrow \begin{bmatrix} 1 & 0 & -1 & 1 \\ 0 & -1 & -1 & 3 \\ 0 & 0 & -6 & 7 \\ 0 & 0 & 0 & 11/3 \end{bmatrix}$, the homogeneous linear system $c_1 M_1 +$

$c_2 M_2 + c_3 M_3 + c_4 M_4 = \begin{bmatrix} 0 & 0 \\ 0 & 0 \end{bmatrix}$ has only the trivial solution and hence, the matrices are linearly independent.

15. Since $\mathbf{v_2} = -\frac{1}{2}\mathbf{v_1}$, the vectors are linearly dependent.

17. Any set of vectors containing the zero vector is linearly dependent.

19. a. Since $\mathbf{A}_2 = -2\mathbf{A}_1$, the column vectors of A are linearly dependent. **b.** Since $\mathbf{A}_3 = \mathbf{A}_1 + \mathbf{A}_2$, the column vectors of A are linearly dependent.

21. Form the matrix with column vectors the three given vectors, that is, let $A = \begin{bmatrix} 1 & -1 & 2 \\ 2 & 0 & a \\ 1 & 1 & 4 \end{bmatrix}$. Since $\det(A) = -2a + 12$, then the vectors are linearly independent if and only if $-2a + 12 \neq 0$, that is $a \neq 6$.

23. a. Since $\begin{vmatrix} 1 & 1 & 1 \\ 1 & 2 & 1 \\ 1 & 3 & 2 \end{vmatrix} = 1$, the vectors are linearly independent.   **b.** Since

$$\left[\begin{array}{ccc|c} 1 & 1 & 1 & 2 \\ 1 & 2 & 1 & 1 \\ 1 & 3 & 2 & 3 \end{array}\right] \longrightarrow \left[\begin{array}{ccc|c} 1 & 0 & 0 & 0 \\ 0 & 1 & 0 & -1 \\ 0 & 0 & 1 & 3 \end{array}\right]$$ the corresponding linear system has the unique solution $(0, -1, 3)$.

Hence $\mathbf{v} = -\mathbf{v_2} + 3\mathbf{v_3}$.

25. Since $\begin{vmatrix} 1 & 2 & 0 \\ -1 & 0 & 3 \\ 2 & 1 & 2 \end{vmatrix} = 13$, the matrix A is invertible, so that $A\mathbf{x} = \mathbf{b}$ has a unique solution for every vector $\mathbf{b}$.

27. Since the equation $c_1(1) + c_2(-2 + 4x^2) + c_3(2x) + c_4(-12x + 8x^3) = 0$, for all x, gives that $c_1 = c_2 = c_3 = c_4 = 0$, the polynomials are linear independent.

29. Since the equation $c_1(2) + c_2(x) + c_3(x^2) + c_4(3x - 1) = 0$, for all x, gives that $c_1 = \frac{1}{2}c_4, c_2 = -3c_4, c_3 = 0, c_4 \in \mathbb{R}$, the polynomials are linearly dependent.

31. In the equation $c_1 \cos \pi x + c_2 \sin \pi x = 0$, if $x = 0$, then $c_1 = 0$, and if $x = \frac{1}{2}$, then $c_2 = 0$. Hence, the functions are linearly independent.

33. In the equation $c_1 x + c_2 x^2 + c_3 e^x = 0$, if $x = 0$, then $c_3 = 0$. Now let $x = 1$, and $x = -1$, which gives the linear system $\begin{cases} c_1 + c_2 & = 0 \\ -c_1 + c_2 & = 0 \end{cases}$. This system has solution $c_1 = 0$ and $c_2 = 0$. Hence the functions are linearly independent.

35. If $\mathbf{u}$ and $\mathbf{v}$ are linearly dependent, then there are scalars a and b, not both 0, such that $a\mathbf{u} + b\mathbf{v} = \mathbf{0}$. If $a \neq 0$, then $\mathbf{u} = -\frac{b}{a}\mathbf{v}$. On the other hand, if there is a scalar c such that $\mathbf{u} = c\mathbf{v}$, then $\mathbf{u} - c\mathbf{v} = \mathbf{0}$.

37. Setting a linear combination of $\mathbf{w_1}, \mathbf{w_2}, \mathbf{w_3}$ to $\mathbf{0}$, we have

$$\mathbf{0} = c_1\mathbf{w_1} + c_2\mathbf{w_2} + c_3\mathbf{w_3} = c_1\mathbf{v_1} + (c_1 + c_2 + c_3)\mathbf{v_2} + (-c_2 + c_3)\mathbf{v_3}.$$

Since the vectors $\mathbf{v_1}, \mathbf{v_2}, \mathbf{v_3}$ are linear independent, then $c_1 = 0, c_1 + c_2 + c_3 = 0$, and $-c_2 + c_3 = 0$. The only solution to this linear system is the trivial solution $c_1 = c_2 = c_3 = 0$, and hence, the vectors $\mathbf{w_1}, \mathbf{w_2}, \mathbf{w_3}$ are linearly independent.

39. Consider $c_1\mathbf{v_1} + c_2\mathbf{v_2} + c_3\mathbf{v_3} = \mathbf{0}$, which is true if and only if $c_3\mathbf{v_3} = -c_1\mathbf{v_1} - c_2\mathbf{v_2}$. If $c_3 \neq 0$, then $\mathbf{v_3}$ would be a linear combination of $\mathbf{v_1}$ and $\mathbf{v_2}$ contradicting the hypothesis that it is not the case. Therefore, $c_3 = 0$. Now since $\mathbf{v_1}$ and $\mathbf{v_2}$ are linearly independent $c_1 = c_2 = 0$.

41. Since $\mathbf{A_1}, \mathbf{A_2}, \ldots, \mathbf{A_n}$ are linearly independent, if

$$A\mathbf{x} = x_1\mathbf{A_1} + \cdots + x_n\mathbf{A_n} = \mathbf{0},$$

then $x_1 = x_2 = \cdots = x_n = 0$. Hence, the only solution to $A\mathbf{x} = \mathbf{0}$ is $\mathbf{x} = \mathbf{0}$.

Review Exercises Chapter 2

1. Since $\begin{vmatrix} a & b \\ c & d \end{vmatrix} = ad - bc \neq 0$, the column vectors are linearly independent. If $ad - bc = 0$, then the column vectors are linearly dependent.

3. The determinant $\begin{vmatrix} a^2 & 0 & 1 \\ 0 & a & 0 \\ 1 & 2 & 1 \end{vmatrix} = a^3 - a = a(a^2 - 1) \neq 0$, if and only if $a \neq \pm 1$, and $a \neq 0$. So the vectors are linearly independent if and only if $a \neq \pm 1$, and $a \neq 0$.

5. a. Since the vectors are not scalar multiples of each other, S is linearly independent.

b. Since

$$\left[\begin{array}{cc|c} 1 & 1 & a \\ 0 & 1 & b \\ 2 & 1 & c \end{array}\right] \rightarrow \left[\begin{array}{cc|c} 1 & 1 & a \\ 0 & 1 & b \\ 0 & 0 & -2a+b+c \end{array}\right],$$

the linear system is inconsistent for any values of a, b and c such that $-2a+b+c \neq 0$. If $a = 1, b = 1, c = 3$,

then the system is inconsistent and $\mathbf{v} = \begin{bmatrix} 1 \\ 1 \\ 3 \end{bmatrix}$ is not a linear combination of the vectors.

c. All vectors $\begin{bmatrix} a \\ b \\ c \end{bmatrix}$ such that $-2a+b+c=0$. **d.** Since $\begin{vmatrix} 1 & 1 & 1 \\ 0 & 1 & 0 \\ 2 & 1 & 0 \end{vmatrix} = -2$, the vectors are linearly

independent. **e.** The augmented matrix of the equation

$$c_1\begin{bmatrix} 1 \\ 0 \\ 2 \end{bmatrix} + c_2\begin{bmatrix} 1 \\ 1 \\ 1 \end{bmatrix} + c_3\begin{bmatrix} 1 \\ 0 \\ 0 \end{bmatrix} = \begin{bmatrix} a \\ b \\ c \end{bmatrix} \quad \text{is} \quad \left[\begin{array}{ccc|c} 1 & 1 & 1 & a \\ 0 & 1 & 0 & b \\ 2 & 1 & 0 & c \end{array}\right] \rightarrow \left[\begin{array}{ccc|c} 1 & 0 & 0 & -\frac{3}{2}b+\frac{1}{2}c \\ 0 & 1 & 0 & b \\ 0 & 0 & 1 & a-\frac{1}{2}b-\frac{1}{2}c \end{array}\right].$$

Hence the system has a unique solution for any values of a, b and c. That is, all vectors in $\mathbb{R}^3$ can be written as a linear combination of the three given vectors.

7. a. Let $A = \begin{bmatrix} 1 & 1 & 2 & 1 \\ -1 & 0 & 1 & 2 \\ 2 & 2 & 0 & 1 \\ 1 & 1 & 2 & 3 \end{bmatrix}$, $\mathbf{x} = \begin{bmatrix} x \\ y \\ z \\ w \end{bmatrix}$, and $\mathbf{b} = \begin{bmatrix} 3 \\ 1 \\ -2 \\ 5 \end{bmatrix}$. **b.** $\det(A) = -8$ **c.** Since the determinant of A is nonzero, the column vectors of A are linearly independent. **d.** Since the determinant of the coefficient matrix is nonzero, the matrix A is invertible, so $A\mathbf{x} = \mathbf{b}$ has a unique solution for every vector **b.** **e.** $x = \frac{11}{4}, y = -\frac{17}{4}, z = \frac{7}{4}, w = 1$

9. a. $x_1\begin{bmatrix} 1 \\ 2 \\ 1 \end{bmatrix} + x_2\begin{bmatrix} 3 \\ -1 \\ 1 \end{bmatrix} + x_3\begin{bmatrix} 2 \\ 3 \\ -1 \end{bmatrix} = \begin{bmatrix} b_1 \\ b_2 \\ b_3 \end{bmatrix}$ **b.** Since $\det(A) = 19$, the linear system $A\mathbf{x} = \mathbf{b}$ has a unique solution for every **b** equal to $\mathbf{x} = A^{-1}\mathbf{b}$. **c.** Since the determinant of A is nonzero, the column vectors of A are linearly independent. **d.** Since the determinant of A is nonzero, we can conclude that the linear system $A\mathbf{x} = \mathbf{b}$ has a unique solution from the fact that A^{-1} exists or from the fact that the column vectors of A are linearly independent.

Chapter Test Chapter 2

1. T

2. F. For example, $\begin{bmatrix} 1 & 0 \\ 0 & 1 \end{bmatrix}$ can not be written as a linear combination of the three matrices.

3. T

4. T

5. F. Since

$$\left[\begin{array}{cc|c} 1 & 2 & 4 \\ 0 & 1 & 3 \\ 1 & 0 & -1 \end{array}\right] \rightarrow \left[\begin{array}{cc|c} 1 & 2 & 4 \\ 0 & 1 & 3 \\ 0 & 0 & 1 \end{array}\right].$$

6. F. Since

$$\left[\begin{array}{cc|c} 1 & 4 & 2 \\ 0 & 3 & 1 \\ 1 & -1 & 0 \end{array}\right] \rightarrow \left[\begin{array}{cc|c} 1 & 4 & 2 \\ 0 & 1 & 0 \\ 0 & 0 & 1 \end{array}\right].$$

7. F. Since

$$\left[\begin{array}{cc|c} 2 & 4 & 1 \\ 1 & 3 & 0 \\ 0 & -1 & 1 \end{array}\right] \rightarrow \left[\begin{array}{cc|c} 2 & 4 & 1 \\ 0 & 1 & -\frac{1}{2} \\ 0 & 0 & \frac{1}{2} \end{array}\right].$$

8. T

9. F. Since $p(x)$ is not a scalar multiple of $q_1(x)$ and any linear combination of $q_1(x)$ and $q_2(x)$ with nonzero scalars will contain an x^2.

10. F. Since there are four vectors in $\mathbb{R}^3$.

11. T

12. T

13. F. The set of all linear combinations of matrices in T is not all 2×2 matrices, but the set of all linear combinations of matrices from S is all 2×2 matrices.

14. T

15. F. Since

$$\begin{vmatrix} s & 1 & 0 \\ 0 & s & 1 \\ 0 & 1 & s \end{vmatrix} = s(s^2 - 1)$$

the vectors are linearly independent if and only if $s \neq 0$ and $s \neq \pm 1$.

16. T

17. T

18. F. Since the column vectors are linearly independent, $\det(A) \neq 0$

19. T

20. T

21. F. If the vector $\mathbf{v_3}$ is a linear combination of $\mathbf{v_1}$ and $\mathbf{v_2}$, then the vectors will be linearly dependent.

22. F. At least one is a linear combination of the others.

23. F. The determinant of the matrix will be zero since the column vectors are linearly dependent.

24. F. The third vector is a combination of the other two and hence, the three together are linearly dependent.

25. T

26. F. An $n \times n$ matrix is invertible if and only if the column vectors are linearly independent.

27. T

28. F. For example, the column vectors of any 3×4 matrix are linearly dependent.

29. T

30. F. The vector can be a linear combination of the linearly independent vectors $\mathbf{v_1}, \mathbf{v_2}$ and $\mathbf{v_3}$.

31. T

32. F. The set of coordinate vectors $\{\mathbf{e_1}, \mathbf{e_2}, \mathbf{e_3}\}$ is linearly independent, but the set $\{\mathbf{e_1}, \mathbf{e_2}, \mathbf{e_3}, \mathbf{e_1} + \mathbf{e_2} + \mathbf{e_3}\}$ is linearly dependent.

33. T

3 Vector Spaces

Exercise Set 3.1

A vector space V is a set with an addition and scalar multiplication defined on the vectors in the set that satisfy the ten axioms. Examples of vector spaces are the Euclidean spaces $\mathbb{R}^n$ with the standard componentwise operations, the set of $m \times n$ matrices $M_{m \times n}$ with the standard componentwise operations, and the set $\mathcal{P}_n$ of polynomials of degree less than or equal to n (including the 0 polynomial) with the standard operations on like terms. To show a set V with an addition and scalar multiplication defined is a vector space requires showing all ten properties hold. To show V is not a vector space it is sufficient to show one of the properties does not hold. The operations defined on a set, even a familiar set, are free to our choosing. For example, on the set $M_{n \times n}$, we can define addition $\oplus$ as matrix multiplication, that is, $A \oplus B = AB$. Then $M_{n \times n}$ is not a vector space since AB may not equal BA, so that $A \oplus B$ is not $B \oplus A$ for all matrices in $M_{n \times n}$. As another example, let $V = \mathbb{R}^3$ and define addition by

$$\begin{bmatrix} x_1 \\ y_1 \\ z_1 \end{bmatrix} \oplus \begin{bmatrix} x_2 \\ y_2 \\ z_2 \end{bmatrix} = \begin{bmatrix} x_1 + x_2 + 1 \\ y_1 + y_2 + 2 \\ z_1 + z_2 - 1 \end{bmatrix}.$$

The additive identity (Axiom (4)) is the unique vector, lets call it $\mathbf{v}_I$, such that $\mathbf{v} \oplus \mathbf{v}_I = \mathbf{v}$ for all vectors $\mathbf{v} \in \mathbb{R}^3$. To determine the additive identity for an arbitrary vector $\mathbf{v}$ we solve the equation $\mathbf{v} \oplus \mathbf{v}_I = \mathbf{v}$, that is,

$$\mathbf{v} \oplus \mathbf{v}_I = \begin{bmatrix} x_1 \\ y_1 \\ z_1 \end{bmatrix} \oplus \begin{bmatrix} x_I \\ y_I \\ z_I \end{bmatrix} = \begin{bmatrix} x_1 + x_I + 1 \\ y_1 + y_I + 2 \\ z_1 + z_I - 1 \end{bmatrix} = \begin{bmatrix} x_1 \\ y_1 \\ z_1 \end{bmatrix} \Leftrightarrow \mathbf{v}_I = \begin{bmatrix} -1 \\ -2 \\ 1 \end{bmatrix}.$$

So the additive identity in this case is not the zero vector, which is the additive identity for the vector space $\mathbb{R}^3$ with the standard operations. Now to find the additive inverse of a vector requires using the additive identity that we just found $\mathbf{v}_I$. So $\mathbf{w}$ is the additive inverse of $\mathbf{v}$ provided $\mathbf{v} \oplus \mathbf{w} = \mathbf{v}_I$, that is,

$$\begin{bmatrix} x_1 \\ y_1 \\ z_1 \end{bmatrix} \oplus \begin{bmatrix} x_2 \\ y_2 \\ z_2 \end{bmatrix} = \begin{bmatrix} x_1 + x_2 + 1 \\ y_1 + y_2 + 2 \\ z_1 + z_2 - 1 \end{bmatrix} = \begin{bmatrix} -1 \\ -2 \\ 1 \end{bmatrix} \Leftrightarrow \mathbf{w} = \begin{bmatrix} -x_1 - 2 \\ -y_1 - 4 \\ -z_1 + 2 \end{bmatrix}.$$

Depending on how a set V and addition and scalar multiplication are defined many of the vector space properties may follow from knowing a vector space that contains V. For example, let $V = \left\{ \begin{bmatrix} x \\ y \\ z \end{bmatrix} \middle| x - y + z = 0 \right\}$ and define addition and scalar multiplication as the standard operations on $\mathbb{R}^3$. It isn't necessary to verify, for example, that if $\mathbf{u}$ and $\mathbf{v}$ are in V, then $\mathbf{u} \oplus \mathbf{v} = \mathbf{v} \oplus \mathbf{u}$, since the vectors are also in $\mathbb{R}^3$ where the property already holds. This applies to most, but not all of the vector space properties. Notice that the vectors in V describe a plane that passes through the origin and hence, is not all of $\mathbb{R}^3$. So to show V is a vector space we would need to show the sum of two vectors from V is another vector in V. In this example, if $\mathbf{u} = \begin{bmatrix} x_1 \\ y_1 \\ z_1 \end{bmatrix}$ and $\mathbf{v} = \begin{bmatrix} x_2 \\ y_2 \\ z_2 \end{bmatrix}$, then $x_1 - y_1 + z_1 = 0$ and $x_2 - y_2 + z_2 = 0$. Then

$$\mathbf{u} \oplus \mathbf{v} = \begin{bmatrix} x_1 \\ y_1 \\ z_1 \end{bmatrix} + \begin{bmatrix} x_2 \\ y_2 \\ z_2 \end{bmatrix} = \begin{bmatrix} x_1 + x_2 \\ y_1 + y_2 \\ z_1 + z_2 \end{bmatrix},$$

and since $(x_1 + x_2) - (y_1 + y_2) + (z_1 + z_2) = (x_1 - y_1 + z_1) + (x_2 - y_2 + z_2) = 0 + 0 = 0$, the sum is also in V. Similarly, $c\mathbf{u}$ is in V for all scalars c. These are the only properties that need to be verified to show that V is a vector space.

■ Solutions to Odd Exercises

1. In order to show that a set V with an addition and scalar multiplication defined is a vector space all ten properties in Definition 1 must be satisfied. To show that V is not a vector space it is sufficient to show any one of the properties does not hold. Since

$$\begin{bmatrix} x_1 \\ y_1 \\ z_1 \end{bmatrix} \oplus \begin{bmatrix} x_2 \\ y_2 \\ z_2 \end{bmatrix} = \begin{bmatrix} x_1 - x_2 \\ y_1 - y_2 \\ z_1 - z_2 \end{bmatrix}$$

and

$$\begin{bmatrix} x_2 \\ y_2 \\ z_2 \end{bmatrix} \oplus \begin{bmatrix} x_1 \\ y_1 \\ z_1 \end{bmatrix} = \begin{bmatrix} x_2 - x_1 \\ y_2 - y_1 \\ z_2 - z_1 \end{bmatrix}$$

do not agree for all pairs of vectors, the operation $\oplus$ is not commutative, so V is not a vector space.

3. The operation $\oplus$ is not associative so V is not a vector space. That is,

$$\left(\begin{bmatrix} x_1 \\ y_1 \\ z_1 \end{bmatrix} \oplus \begin{bmatrix} x_2 \\ y_2 \\ z_2 \end{bmatrix} \right) \oplus \begin{bmatrix} x_3 \\ y_3 \\ z_3 \end{bmatrix} = \begin{bmatrix} 4x_1 + 4x_2 + 2x_3 \\ 4y_1 + 4y_2 + 2y_3 \\ 4z_1 + 4z_2 + 2z_3 \end{bmatrix}$$

and

$$\begin{bmatrix} x_1 \\ y_1 \\ z_1 \end{bmatrix} \oplus \left(\begin{bmatrix} x_2 \\ y_2 \\ z_2 \end{bmatrix} \oplus \begin{bmatrix} x_3 \\ y_3 \\ z_3 \end{bmatrix} \right) = \begin{bmatrix} 2x_1 + 4x_2 + 4x_3 \\ 2y_1 + 4y_2 + 4y_3 \\ 2z_1 + 4z_2 + 4z_3 \end{bmatrix},$$

which do not agree for all vectors.

5.

1. $\begin{bmatrix} x_1 \\ y_1 \end{bmatrix} + \begin{bmatrix} x_2 \\ y_2 \end{bmatrix} = \begin{bmatrix} x_1 + x_2 \\ y_1 + y_2 \end{bmatrix}$ is in $\mathbb{R}^2$.

2. $\begin{bmatrix} x_1 \\ y_1 \end{bmatrix} + \begin{bmatrix} x_2 \\ y_2 \end{bmatrix} = \begin{bmatrix} x_1 + x_2 \\ y_1 + y_2 \end{bmatrix} = \begin{bmatrix} x_2 + x_1 \\ y_2 + y_1 \end{bmatrix} = \begin{bmatrix} x_2 \\ y_2 \end{bmatrix} + \begin{bmatrix} x_1 \\ y_1 \end{bmatrix}$

3. $\left(\begin{bmatrix} x_1 \\ y_1 \end{bmatrix} + \begin{bmatrix} x_2 \\ y_2 \end{bmatrix} \right) + \begin{bmatrix} x_3 \\ y_3 \end{bmatrix}$
$= \begin{bmatrix} x_1 + x_2 + x_3 \\ y_1 + y_2 + y_3 \end{bmatrix}$ and

$\begin{bmatrix} x_1 \\ y_1 \end{bmatrix} + \left(\begin{bmatrix} x_2 \\ y_2 \end{bmatrix} + \begin{bmatrix} x_3 \\ y_3 \end{bmatrix} \right) = \begin{bmatrix} x_1 + x_2 + x_3 \\ y_1 + y_2 + y_3 \end{bmatrix}$

4. Let $\mathbf{0} = \begin{bmatrix} 0 \\ 0 \end{bmatrix}$ and $\mathbf{u} = \begin{bmatrix} x \\ y \end{bmatrix}$. Then $\mathbf{0} + \mathbf{u} = \mathbf{u} + \mathbf{0} = \mathbf{u}$.

5. Let $\mathbf{u} = \begin{bmatrix} x \\ y \end{bmatrix}$ and $-\mathbf{u} = \begin{bmatrix} -x \\ -y \end{bmatrix}$. Then $\mathbf{u} + (-\mathbf{u}) = -\mathbf{u} + \mathbf{u} = \mathbf{0}$.

6. $c \begin{bmatrix} x \\ y \end{bmatrix} = \begin{bmatrix} cx \\ cy \end{bmatrix}$ is a vector in $\mathbb{R}^2$.

7. $c \left(\begin{bmatrix} x_1 \\ y_1 \end{bmatrix} + \begin{bmatrix} x_2 \\ y_2 \end{bmatrix} \right) =$
$c \begin{bmatrix} x_1 + x_2 \\ y_1 + y_2 \end{bmatrix} = \begin{bmatrix} c(x_1 + x_2) \\ c(y_1 + y_2) \end{bmatrix}$
$= \begin{bmatrix} cx_1 + cx_2 \\ cy_1 + cy_2 \end{bmatrix} = c \begin{bmatrix} x_1 \\ y_1 \end{bmatrix} + c \begin{bmatrix} x_2 \\ y_2 \end{bmatrix}$.

8. $(c+d) \begin{bmatrix} x \\ y \end{bmatrix} = \begin{bmatrix} (c+d)x \\ (c+d)y \end{bmatrix} = \begin{bmatrix} cx + dx \\ cy + dy \end{bmatrix} =$
$c \begin{bmatrix} x \\ y \end{bmatrix} + d \begin{bmatrix} x \\ y \end{bmatrix}$.

9. $c\left(d\begin{bmatrix} x \\ y \end{bmatrix}\right) = c\begin{bmatrix} dx \\ dy \end{bmatrix} = \begin{bmatrix} (cd)x \\ (cd)y \end{bmatrix}$

$= (cd)\begin{bmatrix} x \\ y \end{bmatrix}.$

10. $1\begin{bmatrix} x \\ y \end{bmatrix} = \begin{bmatrix} x \\ y \end{bmatrix}.$

7. Since $(c+d)\odot\begin{bmatrix} x \\ y \end{bmatrix} = \begin{bmatrix} x+c+d \\ y \end{bmatrix}$ does not equal

$$c\odot\begin{bmatrix} x \\ y \end{bmatrix} + d\odot\begin{bmatrix} x \\ y \end{bmatrix} = \begin{bmatrix} x+c \\ y \end{bmatrix} + \begin{bmatrix} x+d \\ y \end{bmatrix} = \begin{bmatrix} 2x+c+d \\ 2y \end{bmatrix},$$

for all vectors $\begin{bmatrix} x \\ y \end{bmatrix}$, then V is not a vector space.

9. Since the operation $\oplus$ is not commutative, then V is not a vector space.

11. The zero vector is given by $\mathbf{0} = \begin{bmatrix} 0 \\ 0 \end{bmatrix}$. Since this vector is not in V, then V is not a vector space.

13. a. Since V is not closed under vector addition, then V is not a vector space. That is, if two matrices from V are added, then the row two, column two entry of the sum has the value 2 and hence, the sum is not in V. **b.** Each of the ten vector space axioms are satisfied with vector addition and scalar multiplication defined in this way.

15. The set of upper triangular matrices with the standard componentwise operations is a vector space.

17. The set of invertible matrices is not a vector space. Let $A = I$ and $B = -I$. Then $A + B$ is not invertible, and hence not in V.

19. If A and C are in V and k is a scalar, then $(A + C)B = AB + BC = \mathbf{0}$, and $(kA)B = k(AB) = \mathbf{0}$, so V is closed under addition and scalar multiplication. All the other required properties also hold since V is a subset of the vector space of all matrices with the same operations. Hence, V is a vector space.

21. a. The additive identity is $\mathbf{0} = \begin{bmatrix} 1 & 0 \\ 0 & 1 \end{bmatrix}$. Since $A \oplus A^{-1} = AA^{-1} = I$, then the additive inverse of A is A^{-1}. **b.** If $c = 0$, then cA is not in V. Notice also that addition is not commutative, since AB is not always equal to BA.

23. a. The additive identity is $\mathbf{0} = \begin{bmatrix} 1 \\ 2 \\ 3 \end{bmatrix}$. Let $\mathbf{u} = \begin{bmatrix} 1+a \\ 2-a \\ 3+2a \end{bmatrix}$. Then the additive inverse is $-\mathbf{u} = $

$\begin{bmatrix} 1-a \\ 2+a \\ 3-2a \end{bmatrix}$. **b.** Each of the ten vector space axioms is satisfied. **c.** $0\odot\begin{bmatrix} 1+t \\ 2-t \\ 3+2t \end{bmatrix} = \begin{bmatrix} 1+0t \\ 2-0t \\ 3+2(0)t \end{bmatrix} = \begin{bmatrix} 1 \\ 2 \\ 3 \end{bmatrix}$

25. Each of the ten vector space axioms is satisfied.

27. Each of the ten vector space axioms is satisfied.

29. Since $(f + g)(0) = f(0) + g(0) = 1 + 1 = 2$, then V is not closed under addition and hence is not a vector space.

31. a. The zero vector is given by $f(x+0) = x^3$ and $-f(x + t) = f(x - t)$. **b.** Each of the ten vector space axioms is satisfied.

Exercise Set 3.2

A subset W of a vector space V is a subspace of the vector space if vectors in W, using the same addition and scalar multiplication of V, satisfy the ten vector space properties. That is, W is a vector space. Many of the vector space properties are inherited by W from V. For example, if $\mathbf{u}$ and $\mathbf{v}$ are vectors in W, then they are also vectors in V, so that $\mathbf{u} \oplus \mathbf{v} = \mathbf{v} \oplus \mathbf{u}$. On the other hand, the additive identity may not be a vector in W, which is a requirement for being a vector space. To show that a subset is a subspace it is sufficient to verify that

if $\mathbf{u}$ and $\mathbf{v}$ are in W and c is a scalar, then $\mathbf{u} + c\mathbf{v}$ is another vector in W.

For example, let

$$W = \left\{ \left. \begin{bmatrix} s - 2t \\ t \\ s + t \end{bmatrix} \right| s, t \in \mathbb{R} \right\},$$

which is a subset of $\mathbb{R}^3$. Notice that if $s = t = 0$, then the additive identity $\begin{bmatrix} 0 \\ 0 \\ 0 \end{bmatrix}$ for the vector space $\mathbb{R}^3$ is

also in W. Let $\mathbf{u} = \begin{bmatrix} s - 2t \\ t \\ s + t \end{bmatrix}$, and $\mathbf{v} = \begin{bmatrix} a - 2b \\ b \\ a + b \end{bmatrix}$ denote two arbitrary vectors in W. Notice that we have

to use different parameters for the two vectors since the vectors may be different. Next we form the linear combination

$$\mathbf{u} + c\mathbf{v} = \begin{bmatrix} s - 2t \\ t \\ s + t \end{bmatrix} + c \begin{bmatrix} a - 2b \\ b \\ a + b \end{bmatrix}$$

and simplify the sum to one vector. So

$$\mathbf{u} + c\mathbf{v} = \begin{bmatrix} s - 2t \\ t \\ s + t \end{bmatrix} + c \begin{bmatrix} a - 2b \\ b \\ a + b \end{bmatrix} = \begin{bmatrix} (s - 2t) + (a - 2b) \\ t + b \\ (s + t) + (a + b) \end{bmatrix}$$

but this is not sufficient to show the vector is in W since the vector must be written in terms of just two parameters in the locations described in the definition of W. Continuing the simplification, we have that

$$\mathbf{u} + c\mathbf{v} = \begin{bmatrix} (s + a) - 2(t + b) \\ t + b \\ (s + a) + (t + b) \end{bmatrix}$$

and now the vector $\mathbf{u} + c\mathbf{v}$ is in the required form with two parameters $s + a$ and $t + b$. Hence, W is a subspace of $\mathbb{R}^3$. An arbitrary vector in W can also be written as

$$\begin{bmatrix} s - 2t \\ t \\ s + t \end{bmatrix} = s \begin{bmatrix} 1 \\ 0 \\ 1 \end{bmatrix} + t \begin{bmatrix} -2 \\ 1 \\ 1 \end{bmatrix}$$

and in this case $\begin{bmatrix} 1 \\ 0 \\ 1 \end{bmatrix}$ and $\begin{bmatrix} -2 \\ 1 \\ 1 \end{bmatrix}$ are linearly independent, so W is a plane in $\mathbb{R}^3$. The set W consists of

all linear combinations of the two vectors $\begin{bmatrix} 1 \\ 0 \\ 1 \end{bmatrix}$ and $\begin{bmatrix} -2 \\ 1 \\ 1 \end{bmatrix}$, called the span of the vectors and written

$$W = \mathbf{span} \left\{ \begin{bmatrix} 1 \\ 0 \\ 1 \end{bmatrix}, \begin{bmatrix} -2 \\ 1 \\ 1 \end{bmatrix} \right\}.$$

The span of a set of vectors is always a subspace. Important facts to keep in mind are:

- There are linearly independent vectors that span a vector space. The coordinate vectors of $\mathbb{R}^3$ are a simple example.

- Two linearly independent vectors can not span $\mathbb{R}^3$, since they describe a plane and one vector can not span $\mathbb{R}^2$ since all linear combinations describe a line.

- Two linearly dependent vectors can not span $\mathbb{R}^2$. Let $S = \mathbf{span}\left\{ \begin{bmatrix} 2 \\ 1 \end{bmatrix}, \begin{bmatrix} -4 \\ -2 \end{bmatrix} \right\}$. If $\mathbf{v}$ in in S, then there are scalars c_1 and c_2 such that

$$\mathbf{v} = c_1 \begin{bmatrix} 2 \\ 1 \end{bmatrix} + c_2 \begin{bmatrix} -4 \\ -2 \end{bmatrix} = c_1 \begin{bmatrix} 2 \\ 1 \end{bmatrix} + c_2 \left(-2 \begin{bmatrix} 2 \\ 1 \end{bmatrix} \right) = (c_1 - 2c_2) \begin{bmatrix} 2 \\ 1 \end{bmatrix}$$

and hence every vector in the span of S is a linear combination of only one vector.

- A linearly dependent set of vectors can span a vector space. For example, let $S = \left\{ \begin{bmatrix} 1 \\ 0 \end{bmatrix}, \begin{bmatrix} 0 \\ 1 \end{bmatrix}, \begin{bmatrix} 2 \\ 3 \end{bmatrix} \right\}$. Since the coordinate vectors are in S, then $\mathbf{span}(S) = \mathbb{R}^2$, but the vectors are linearly dependent since $\begin{bmatrix} 2 \\ 3 \end{bmatrix} = 2 \begin{bmatrix} 1 \\ 0 \end{bmatrix} + 3 \begin{bmatrix} 0 \\ 1 \end{bmatrix}$.

In general, to determine whether or not a vector $\mathbf{v} = \begin{bmatrix} v_1 \\ v_2 \\ \vdots \\ v_n \end{bmatrix}$ is in $\mathbf{span}\{\mathbf{u_1}, \ldots, \mathbf{u_k}\}$, start with the vector equation

$$c_1 \mathbf{u_1} + c_2 \mathbf{u_2} + \cdots + c_k \mathbf{u_k} = \mathbf{v},$$

and then solve the resulting linear system. These ideas apply to all vector spaces not just the Euclidean spaces. For example, if $S = \{ A \in M_{2 \times 2} \mid A \text{ is invertible} \}$, then S is not a subspace of the vector space of all 2×2 matrices. For example, the matrices $\begin{bmatrix} 1 & 0 \\ 0 & -1 \end{bmatrix}$ and $\begin{bmatrix} 1 & 0 \\ 0 & 1 \end{bmatrix}$ are both invertible, so are in S, but $\begin{bmatrix} 1 & 0 \\ 0 & -1 \end{bmatrix} + \begin{bmatrix} 1 & 0 \\ 0 & 1 \end{bmatrix} = \begin{bmatrix} 2 & 0 \\ 0 & 0 \end{bmatrix}$, which is not invertible. To determine whether of not $\begin{bmatrix} 3 & -1 \\ 1 & 1 \end{bmatrix}$ is in the span of the two matrices $\begin{bmatrix} 1 & 2 \\ 0 & 1 \end{bmatrix}$ and $\begin{bmatrix} -1 & 0 \\ 1 & 1 \end{bmatrix}$, start with the equation

$$c_1 \begin{bmatrix} 1 & 2 \\ 0 & 1 \end{bmatrix} + c_2 \begin{bmatrix} -1 & 0 \\ 1 & 1 \end{bmatrix} = \begin{bmatrix} 2 & -1 \\ 1 & 1 \end{bmatrix} \Leftrightarrow \begin{bmatrix} c_1 - c_2 & 2c_1 \\ c_2 & c_1 + c_2 \end{bmatrix} = \begin{bmatrix} 2 & -1 \\ 1 & 1 \end{bmatrix}.$$

The resulting linear system is $c_1 - c_2 = 2, 2c_1 = -1, c_2 = 1, c_1 + c_2 = 1$, is inconsistent and hence, the matrix is not in the span of the other two matrices.

■ Solutions to Odd Exercises

1. Let $\begin{bmatrix} 0 \\ y_1 \end{bmatrix}$ and $\begin{bmatrix} 0 \\ y_2 \end{bmatrix}$ be two vectors in S and c a scalar. Then $\begin{bmatrix} 0 \\ y_1 \end{bmatrix} + c \begin{bmatrix} 0 \\ y_2 \end{bmatrix} = \begin{bmatrix} 0 \\ y_1 + cy_2 \end{bmatrix}$ is in S, so S is a subspace of $\mathbb{R}^2$.

3. The set S is not a subspace of $\mathbb{R}^2$. If $\mathbf{u} = \begin{bmatrix} 2 \\ -1 \end{bmatrix}, \mathbf{v} = \begin{bmatrix} -1 \\ 3 \end{bmatrix}$, then $\mathbf{u} + \mathbf{v} = \begin{bmatrix} 1 \\ 2 \end{bmatrix} \notin S$.

5. The set S is not a subspace of $\mathbb{R}^2$. If $\mathbf{u} = \begin{bmatrix} 0 \\ -1 \end{bmatrix}$ and $c = 0$, then $c\mathbf{v} = \begin{bmatrix} 0 \\ 0 \end{bmatrix} \notin S$.

7. Since

$$\begin{bmatrix} x_1 \\ x_2 \\ x_3 \end{bmatrix} + c \begin{bmatrix} y_1 \\ y_2 \\ y_3 \end{bmatrix} = \begin{bmatrix} x_1 + cy_1 \\ x_2 + cy_2 \\ x_3 + cy_3 \end{bmatrix}$$

and $(x_1 + cy_1) + (x_3 + cy_3) = (x_1 + x_3) + c(y_1 + y_3) = -2(c+1) = 2$ if and only if $c = -2$, S is not a subspace of $\mathbb{R}^3$.

9. Since for all real numbers s, t, c, we have that

$$\begin{bmatrix} s - 2t \\ s \\ t + s \end{bmatrix} + c \begin{bmatrix} x - 2y \\ x \\ y + x \end{bmatrix} = \begin{bmatrix} (s + cx) - 2(t + cy) \\ s + cx \\ (t + cy) + (s + cx) \end{bmatrix},$$

is in S, then S is a subspace.

11. If A and B are symmetric matrices and c is a scalar, then $(A + cB)^t = A^t + cB^t = A + cB$, so S is a subspace.

13. Since the sum of invertible matrices may not be invertible, S is not a subspace.

15. If A and B are upper triangular matrices, then $A + B$ and cB are also upper triangular, so S is a subspace.

17. Yes, S is a subspace.

19. No, S is not a subspace since $x^3 - x^3 = 0$, which is not a polynomial of degree 3.

21. If $p(x)$ and $q(x)$ are polynomials with $p(0) = 0$ and $q(0) = 0$, then

$$(p + q)(0) = p(0) + q(0) = 0,$$

and

$$(cq)(0) = cq(0) = 0,$$

so S is a subspace.

23. The set S is not a subspace, since for example $(2x^2 + 1) - (x^2 + 1) = x^2$, which is not in S.

25. The vector $\mathbf{v}$ is in the span of $S = \{\mathbf{v_1}, \mathbf{v_2}, \mathbf{v_3}\}$ provided there are scalars c_1, c_2, and c_3 such that $\mathbf{v} = c_1 \mathbf{v_1} + c_2 \mathbf{v_2} + c_3 \mathbf{v_3}$. Row reduce the augmented matrix $[\mathbf{v_1}\ \mathbf{v_2}\ \mathbf{v_3}\ |\ \mathbf{v}]$. We have that

$$\left[\begin{array}{ccc|c} 1 & -1 & -1 & 1 \\ 1 & -1 & 2 & -1 \\ 0 & 1 & 0 & 1 \end{array}\right] \rightarrow \left[\begin{array}{ccc|c} 1 & -1 & -1 & 1 \\ 0 & 1 & 0 & 1 \\ 0 & 0 & 3 & -2 \end{array}\right],$$

and since the linear system has a (unique) solution, the vector $\mathbf{v}$ is in the span.

27. Since

$$\left[\begin{array}{ccc|c} 1 & 0 & 1 & -2 \\ 1 & 1 & -1 & 1 \\ 0 & 2 & -4 & 6 \\ -1 & 1 & -3 & 5 \end{array}\right] \rightarrow \left[\begin{array}{ccc|c} 1 & 0 & 1 & -2 \\ 0 & 1 & -2 & 3 \\ 0 & 0 & 0 & 0 \\ 0 & 0 & 0 & 0 \end{array}\right], \quad \text{the vector } \mathbf{v} \text{ is in the span.}$$

29. Since

$$c_1(1 + x) + c_2(x^2 - 2) + c_3(3x) = 2x^2 - 6x - 11 \text{ if and only if } (c_1 - 2c_2) + (c_1 + 3c_3)x + c_2x^2 = 2x^2 - 6x - 11,$$

we have that $c_1 = -7, c_2 = 2, c_3 = \frac{1}{3}$ and hence, the polynomial is in the span.

31. Row reducing

$$\left[\begin{array}{cc|c} 2 & 1 & a \\ -1 & 3 & b \\ -2 & -1 & c \end{array}\right] \longrightarrow \left[\begin{array}{cc|c} -1 & 3 & b \\ 0 & 7 & a + 2b \\ 0 & 0 & a + c \end{array}\right], \quad \text{so } \mathbf{span}(S) = \left\{ \begin{bmatrix} a \\ b \\ c \end{bmatrix} \middle| \ a + c = 0 \right\}.$$

33. The equation $c_1 \begin{bmatrix} 1 & 2 \\ 1 & 0 \end{bmatrix} + c_2 \begin{bmatrix} 1 & -1 \\ 0 & 1 \end{bmatrix} = \begin{bmatrix} a & b \\ c & d \end{bmatrix}$, leads to the linear system $c_1 + c_2 = a, 2c_1 - c_2 = b, c_1 = c, c_2 = d$, which gives

$$\mathbf{span}(S) = \left\{ \begin{bmatrix} a & b \\ \frac{a+b}{3} & \frac{2a-b}{3} \end{bmatrix} \middle| \ a, b \in \mathbb{R} \right\}.$$

35.

$$\mathbf{span}(S) = \left\{ ax^2 + bx + c \mid a - c = 0 \right\}.$$

37. a.

$$\mathbf{span}(S) = \left\{ \begin{bmatrix} a \\ b \\ \frac{b-2a}{3} \end{bmatrix} \,\middle|\, a, b \in \mathbb{R} \right\}.$$

b. The set S is linearly independent.

39. a. $\mathbf{span}(S) = \mathbb{R}^3$
b. Yes, S is linearly independent.

41. a. $\mathbf{span}(S) = \mathbb{R}^3$
b. The set S is linearly dependent.
c. $\mathbf{span}(T) = \mathbb{R}^3$; T is linearly dependent.
d. $\mathbf{span}(H) = \mathbb{R}^3$; H is linearly independent.

43. a. $\mathbf{span}(S) = \mathcal{P}_2$ **b.** No, S is linearly dependent. **c.** $2x^2 + 3x + 5 = 2(1) - (x - 3) + 2(x^2 + 2x)$ **d.** T is linearly independent; $\mathbf{span}(T) = \mathcal{P}_3$

45. a., b. Since

$$\begin{bmatrix} -s \\ s - 5t \\ 2s + 3t \end{bmatrix} = s \begin{bmatrix} -1 \\ 1 \\ 2 \end{bmatrix} + t \begin{bmatrix} 0 \\ -5 \\ 3 \end{bmatrix}, \text{ then } S = \mathbf{span} \left\{ \begin{bmatrix} -1 \\ 1 \\ 2 \end{bmatrix}, \begin{bmatrix} 0 \\ -5 \\ 3 \end{bmatrix} \right\}.$$

Therefore, S is a subspace. **c.** The vectors found in part (b) are linearly independent. **d.** Since the span of two linearly independent vectors in $\mathbb{R}^3$ is a plane, then $S \neq \mathbb{R}^3$

47. Since $A(\mathbf{x} + c\mathbf{y}) = \begin{bmatrix} 1 \\ 2 \end{bmatrix} + c \begin{bmatrix} 1 \\ 2 \end{bmatrix} = \begin{bmatrix} 1 \\ 2 \end{bmatrix}$ if and only if $c = 0$, then S is not a subspace.

49. Let $B_1, B_2 \in S$. Since A commutes with B_1 and B_2, we have that

$$A(B_1 + cB_2) = AB_1 + cAB_2 = B_1 A + c(B_2 A) = (B_1 + cB_2)A$$

and hence, $B_1 + cB_2 \in S$ and S is a subspace.

51. Let $\mathbf{w} \in S + T$, so that $\mathbf{w} = \mathbf{u} + \mathbf{v}$, where $\mathbf{u} \in S$, and $\mathbf{v} \in T$. Then there are scalars $c_1, \ldots, c_m$ and $d_1, \ldots, d_n$ such that $\mathbf{w} = \sum_{i=1}^{m} c_i \mathbf{u_i} + \sum_{i=1}^{n} d_i \mathbf{v_i}$. Therefore, $\mathbf{w} \in \mathbf{span}\{\mathbf{u_1}, \ldots, \mathbf{u_m}, \mathbf{v_1}, \ldots, \mathbf{v_n}\}$ and we have shown that $S + T \subseteq \mathbf{span}\{\mathbf{u_1}, \ldots, \mathbf{u_m}, \mathbf{v_1}, \ldots, \mathbf{v_n}\}$.

Now let $\mathbf{w} \in \mathbf{span}\{\mathbf{u_1}, \ldots, \mathbf{u_m}, \mathbf{v_1}, \ldots, \mathbf{v_n}\}$, so there are scalars $c_1, \ldots, c_{m+n}$ such that $\mathbf{w} = c_1 \mathbf{u_1} + \cdots + c_m \mathbf{u_m} + c_{m+1} \mathbf{v_1} + \cdots + c_{m+n} \mathbf{v_n}$, which is in $S + T$. Therefore, $\mathbf{span}\{\mathbf{u_1}, \ldots, \mathbf{u_m}, \mathbf{v_1}, \ldots, \mathbf{v_n}\} \subseteq S + T$.

Exercise Set 3.3

In Section 3.3 of the text, the connection between a spanning set of a vector space and linear independence is completed. The minimal spanning sets, minimal in the sense of the number of vectors in the set, are those that are linear independent. A basis for a vector space V is a set B such that B is linearly independent and $\mathbf{span}(B) = V$. For example,

- $B = \{\mathbf{e_1}, \mathbf{e_2}, \ldots, \mathbf{e_n}\}$ is a basis for $\mathbb{R}^n$

- $B = \left\{ \begin{bmatrix} 1 & 0 \\ 0 & 0 \end{bmatrix}, \begin{bmatrix} 0 & 1 \\ 0 & 0 \end{bmatrix}, \begin{bmatrix} 0 & 0 \\ 1 & 0 \end{bmatrix}, \begin{bmatrix} 0 & 0 \\ 0 & 1 \end{bmatrix} \right\}$ is a basis for $M_{2 \times 2}$

- $B = \{1, x, x^2, \ldots, x^n\}$ is a basis for $\mathcal{P}_n$.

Every vector space has infinitely many bases. For example, if $c \neq 0$, then $B = \{c\mathbf{e_1}, \mathbf{e_2}, \ldots, \mathbf{e_n}\}$ is another basis for $\mathbb{R}^n$. But all bases for a vector space have the same number of vectors, called the dimension of the vector space, and denoted by $\dim(V)$. As a consequence of the bases noted above:

- $\dim(\mathbb{R}^n) = n$

- $\dim(M_{2 \times 2}) = 4$ and in general $\dim(M_{m \times n}) = mn$

- $\dim(\mathcal{P}_n) = n + 1$.

If $S = \{\mathbf{v_1}, \mathbf{v_2}, \ldots, \mathbf{v_m}\}$ is a subset of a vector space V and $\dim(V) = n$ recognizing the following possibilities will be useful in the exercise set:

- If the number of vectors in S exceeds the dimension of V, that is, $m > n$, then S is linearly dependent and hence, can not be a basis.

- If $m > n$, then $\mathbf{span}(S)$ can equal V, but in this case some of the vectors are linear combinations of others and the set S can be trimmed down to a basis for V.

- If $m \leq n$, then the set S can be either linearly independent of linearly dependent.

- If $m < n$, then S can not be a basis for V, since in this case $\mathbf{span}(S) \neq V$.

- If $m < n$ and the vectors in S are linearly independent, then S can be expanded to a basis for V.

- If $m = n$, then S will be a basis for V if either S is linearly independent or $\mathbf{span}(S) = V$. So in this case it is sufficient to verify only one of the conditions.

The two vectors $\mathbf{v_1} = \begin{bmatrix} 1 \\ -1 \\ 2 \end{bmatrix}$ and $\mathbf{v_2} = \begin{bmatrix} 3 \\ -1 \\ 2 \end{bmatrix}$ are linearly independent but can not be a basis for $\mathbb{R}^3$ since all bases for $\mathbb{R}^3$ must have three vectors. To expand to a basis start with the matrix

$$\begin{bmatrix} 1 & 3 & 1 & 0 & 0 \\ -1 & -1 & 0 & 1 & 0 \\ 2 & 2 & 0 & 0 & 1 \end{bmatrix} \xrightarrow{\text{reduce to the echelon form}} \begin{bmatrix} 1 & 3 & 1 & 0 & 0 \\ 0 & 2 & 1 & 1 & 0 \\ 0 & 0 & 0 & 2 & 1 \end{bmatrix}.$$

The pivots in the echelon form matrix are located in columns one, two and four, so the corresponding column vectors in the original matrix form the basis. So

$$B = \left\{ \begin{bmatrix} 1 \\ -1 \\ 2 \end{bmatrix}, \begin{bmatrix} 3 \\ -1 \\ 2 \end{bmatrix}, \begin{bmatrix} 0 \\ 1 \\ 0 \end{bmatrix} \right\} \text{ is a basis for } \mathbb{R}^3.$$

To trim a set of vectors that span the space to a basis the procedure is the same. For example, the set

$$S = \left\{ \begin{bmatrix} 0 \\ -1 \\ -1 \end{bmatrix}, \begin{bmatrix} 2 \\ 2 \\ 1 \end{bmatrix}, \begin{bmatrix} 0 \\ 2 \\ 2 \end{bmatrix}, \begin{bmatrix} 3 \\ -1 \\ -1 \end{bmatrix} \right\}$$

is not a basis for $\mathbb{R}^3$ since there are four vectors and hence, S is linearly dependent. Since

$$\begin{bmatrix} 0 & 2 & 0 & 3 \\ -1 & 2 & 2 & -1 \\ -1 & 1 & 2 & -1 \end{bmatrix} \xrightarrow{\text{reduces to}} \begin{bmatrix} -1 & 2 & 2 & -1 \\ 0 & 2 & 0 & 3 \\ 0 & 0 & 0 & 3 \end{bmatrix},$$

then the span of the four vectors is $\mathbb{R}^3$. The pivots in the reduced matrix are in columns one, two and four, so a basis for $\mathbb{R}^3$ is $\left\{ \begin{bmatrix} 0 \\ -1 \\ -1 \end{bmatrix}, \begin{bmatrix} 2 \\ 2 \\ 1 \end{bmatrix}, \begin{bmatrix} 3 \\ -1 \\ -1 \end{bmatrix} \right\}$.

■ Solutions to Odd Exercises

1. Since $\dim(\mathbb{R}^3) = 3$ every basis for $\mathbb{R}^3$ has three vectors. Therefore, since S has only two vectors it is not a basis for $\mathbb{R}^3$.

3. Since the third vector can be written as the sum of the first two, the set S is linearly dependent and hence, is not a a basis for $\mathbb{R}^3$.

5. Since the third polynomial is a linear combination of the first two, the set S is linearly dependent and hence is not a basis for $\mathcal{P}_3$.

7. The two vectors in S are not scalar multiples and hence, the set S is linearly independent. Since every linearly independent set of two vectors in $\mathbb{R}^2$ is a basis, the set S is a basis.

9. Since

$$\begin{bmatrix} 1 & 0 & 0 \\ -1 & -2 & 2 \\ 1 & -3 & -2 \end{bmatrix} \xrightarrow{\text{reduces to}} \begin{bmatrix} 1 & 0 & 0 \\ 0 & -2 & 2 \\ 0 & 0 & -5 \end{bmatrix},$$

S is a linearly independent set of three vectors in $\mathbb{R}^3$ and hence, S is a basis.

11. Since

$$\begin{bmatrix} 1 & 1 & 0 & 1 \\ 0 & 1 & 1 & 0 \\ 1 & -1 & -1 & 0 \\ 0 & 0 & 2 & 1 \end{bmatrix} \xrightarrow{\text{reduces to}} \begin{bmatrix} 1 & 1 & 0 & 1 \\ 0 & 1 & 1 & 0 \\ 0 & 0 & 1 & -1 \\ 0 & 0 & 0 & 3 \end{bmatrix},$$

the set S is a linearly independent set of four matrices in $M_{2\times2}$. Since $\dim(M_{2\times2}) = 4$, then S is a basis.

13. Since

$$\begin{bmatrix} -1 & 1 & 1 \\ 2 & 0 & 1 \\ 1 & 1 & 1 \end{bmatrix} \xrightarrow{\text{reduces to}} \begin{bmatrix} -1 & 1 & 1 \\ 0 & 2 & 3 \\ 0 & 0 & -1 \end{bmatrix},$$

the set S is a linearly independent set of three vectors in $\mathbb{R}^3$, so is a basis.

15. Since

$$\begin{bmatrix} 1 & 2 & 2 & -1 \\ 1 & 1 & 4 & 2 \\ -1 & 3 & 2 & 0 \\ 1 & 1 & 5 & 3 \end{bmatrix} \xrightarrow{\text{reduces to}} \begin{bmatrix} 1 & 2 & 2 & -1 \\ 0 & -1 & 2 & 3 \\ 0 & 0 & 1 & 1 \\ 0 & 0 & 0 & 0 \end{bmatrix},$$

the homogeneous linear system has infinitely many solutions so the set S is linearly dependent and is therefore not a basis for $\mathbb{R}^4$.

17. Notice that $\frac{1}{3}(2x^2 + x + 2 + 2(-x^2 + x) - 2(1)) = x$ and $2x^2 + x + 2 + (-x^2 + x) - 2x - 2 = x^2$, so the span of S is $\mathcal{P}_2$. Since $\dim(\mathcal{P}_2) = 3$, the set S is a basis.

19. Every vector in S can be written as $\begin{bmatrix} s + 2t \\ -s + t \\ t \end{bmatrix} = s\begin{bmatrix} 1 \\ -1 \\ 0 \end{bmatrix} + t\begin{bmatrix} 2 \\ 1 \\ 1 \end{bmatrix}$. Since the vectors $\begin{bmatrix} 1 \\ -1 \\ 0 \end{bmatrix}$ and $\begin{bmatrix} 2 \\ 1 \\ 1 \end{bmatrix}$ are linear independent a basis for S is $B = \left\{ \begin{bmatrix} 1 \\ -1 \\ 0 \end{bmatrix}, \begin{bmatrix} 2 \\ 1 \\ 1 \end{bmatrix} \right\}$ and $\dim(S) = 2$.

21. Every 2×2 symmetric matrix has the form $\begin{bmatrix} a & b \\ b & d \end{bmatrix} = a\begin{bmatrix} 1 & 0 \\ 0 & 0 \end{bmatrix} + b\begin{bmatrix} 0 & 1 \\ 1 & 0 \end{bmatrix} + c\begin{bmatrix} 0 & 0 \\ 0 & 1 \end{bmatrix}$. Since the three matrices on the right are linearly independent a basis for S is $B = \left\{ \begin{bmatrix} 1 & 0 \\ 0 & 0 \end{bmatrix}, \begin{bmatrix} 0 & 1 \\ 1 & 0 \end{bmatrix}, \begin{bmatrix} 0 & 0 \\ 0 & 1 \end{bmatrix} \right\}$ and $\dim(S) = 3$.

23. Since every polynomial $p(x)$ in S satisfies $p(0) = 0$, we have that $p(x) = ax + bx^2$. Therefore, a basis for S is $B = \{x, x^2\}$ and $\dim(S) = 2$.

25. The set S is already a basis for $\mathbb{R}^3$ since it is a linearly independent set of three vectors in $\mathbb{R}^3$.

27. The vectors can not be a basis since a set of four vectors in $\mathbb{R}^3$ is linearly dependent. To trim the set down to a basis for the span row reduce the matrix with column vectors the vectors in S. This gives

$$\begin{bmatrix} 2 & 0 & -1 & 2 \\ -3 & 2 & -1 & 3 \\ 0 & 2 & 0 & -1 \end{bmatrix} \longrightarrow \begin{bmatrix} 2 & 0 & -1 & 2 \\ 0 & 2 & -\frac{5}{2} & 6 \\ 0 & 0 & \frac{5}{2} & -7 \end{bmatrix}.$$

A basis for the span consists of the column vectors in the original matrix corresponding to the pivot columns of the row echelon matrix. So a basis for the span of S is given by

$$B = \left\{ \begin{bmatrix} 2 \\ -3 \\ 0 \end{bmatrix}, \begin{bmatrix} 0 \\ 2 \\ 2 \end{bmatrix}, \begin{bmatrix} -1 \\ -1 \\ 0 \end{bmatrix} \right\}.$$ Observe that $\mathbf{span}(S) = \mathbb{R}^3$.

29. The vectors can not be a basis since a set of four vectors in $\mathbb{R}^3$ is linearly dependent. To trim the set down to a basis for the span row reduce the matrix with column vectors the vectors in S. This gives

$$\begin{bmatrix} 2 & 0 & 2 & 4 \\ -3 & 2 & -1 & 0 \\ 0 & 2 & 2 & 4 \end{bmatrix} \longrightarrow \begin{bmatrix} 2 & 0 & 2 & 4 \\ 0 & 2 & 2 & 6 \\ 0 & 0 & 0 & -2 \end{bmatrix}.$$

A basis for the span consists of the column vectors in the original matrix corresponding to the pivot columns of the row echelon matrix. So a basis for the span of S is given by

$$B = \left\{ \begin{bmatrix} 2 \\ -3 \\ 0 \end{bmatrix}, \begin{bmatrix} 0 \\ 2 \\ 2 \end{bmatrix}, \begin{bmatrix} 4 \\ 0 \\ 4 \end{bmatrix} \right\}.$$ Observe that $\mathbf{span}(S) = \mathbb{R}^3$.

31. Form the 3×5 matrix with first two column vectors the vectors in S and then augment the identity matrix. Reducing this matrix, we have that

$$\begin{bmatrix} 2 & 1 & 1 & 0 & 0 \\ -1 & 0 & 0 & 1 & 0 \\ 3 & 2 & 0 & 0 & 1 \end{bmatrix} \longrightarrow \begin{bmatrix} 2 & 1 & 1 & 0 & 0 \\ 0 & 1 & 1 & 2 & 0 \\ 0 & 0 & -2 & -1 & 1 \end{bmatrix}.$$

A basis for $\mathbb{R}^3$ consists of the column vectors in the original matrix corresponding to the pivot columns of the row echelon matrix. So a basis for $\mathbb{R}^3$ containing S is $B = \left\{ \begin{bmatrix} 2 \\ -1 \\ 3 \end{bmatrix}, \begin{bmatrix} 1 \\ 0 \\ 2 \end{bmatrix}, \begin{bmatrix} 1 \\ 0 \\ 0 \end{bmatrix} \right\}.$

33. A basis for $\mathbb{R}^4$ containing S is

$$B = \left\{ \begin{bmatrix} 1 \\ -1 \\ 2 \\ 4 \end{bmatrix}, \begin{bmatrix} 3 \\ 1 \\ 1 \\ 2 \end{bmatrix}, \begin{bmatrix} 1 \\ 0 \\ 0 \\ 0 \end{bmatrix}, \begin{bmatrix} 0 \\ 0 \\ 1 \\ 0 \end{bmatrix} \right\}.$$

35. A basis for $\mathbb{R}^3$ containing S is

$$B = \left\{ \begin{bmatrix} -1 \\ 1 \\ 3 \end{bmatrix}, \begin{bmatrix} 1 \\ 1 \\ 1 \end{bmatrix}, \begin{bmatrix} 1 \\ 0 \\ 0 \end{bmatrix} \right\}.$$

37. Let $\mathbf{e_{ii}}$ denote the $n \times n$ matrix with a 1 in the row i, column i component and 0 in all other locations. Then $B = \{\mathbf{e_{ii}} \mid 1 \le i \le n\}$ is a basis for the subspace of all $n \times n$ diagonal matrices.

39. It is sufficient to show that the set S' is linearly independent. Consider the equation $c_1 A\mathbf{v_1} + c_2 A\mathbf{v_2} + \cdots + c_n A\mathbf{v_n} = \mathbf{0}$. This is equivalent to $A(c_1\mathbf{v_1} + c_2\mathbf{v_2} + \cdots + c_n\mathbf{v_n}) = \mathbf{0}$. Multiplying both sides of this equation by A^{-1} gives the equation $c_1\mathbf{v_1} + c_2\mathbf{v_2} + \cdots + c_n\mathbf{v_n} = \mathbf{0}$. Since S is linearly independent, then $c_1 = c_2 = \cdots c_n = 0$, so that S' is linearly independent.

41. Since H is a subspace of V, then $H \subseteq V$. Let $S = \{\mathbf{v_1}, \mathbf{v_2}, \ldots, \mathbf{v_n}\}$ be a basis for H, so that S is a linearly independent set of vectors in V. Since $\dim(V) = n$, then S is also a basis for V. Now let $\mathbf{v}$ be a vector in V. Since S is a basis for V, then there exist scalars c_1, c_2, $\ldots c_n$ such that $c_1\mathbf{v_1} + c_2\mathbf{v_2} + \cdots + c_n\mathbf{v_n} = \mathbf{v}$. Since S is a basis for H, then $\mathbf{v}$ is a linear combination of vectors in H and is therefore, also in H. Hence, $V \subseteq H$ and we have that $H = V$.

43. Every vector in W can be written as a linear combination of the form

$$\begin{bmatrix} 2s + t + 3r \\ 3s - t + 2r \\ s + t + 2r \end{bmatrix} = s\begin{bmatrix} 2 \\ 3 \\ 1 \end{bmatrix} + t\begin{bmatrix} 1 \\ -1 \\ 1 \end{bmatrix} + r\begin{bmatrix} 3 \\ 2 \\ 2 \end{bmatrix}.$$

But $\begin{bmatrix} 3 \\ 2 \\ 2 \end{bmatrix} = \begin{bmatrix} 2 \\ 3 \\ 1 \end{bmatrix} + \begin{bmatrix} 1 \\ -1 \\ 1 \end{bmatrix}$ and hence, $\mathbf{span}\left\{ \begin{bmatrix} 2 \\ 3 \\ 1 \end{bmatrix}, \begin{bmatrix} 1 \\ -1 \\ 1 \end{bmatrix} \right\} = \mathbf{span}\left\{ \begin{bmatrix} 2 \\ 3 \\ 1 \end{bmatrix}, \begin{bmatrix} 1 \\ -1 \\ 1 \end{bmatrix}, \begin{bmatrix} 3 \\ 2 \\ 2 \end{bmatrix} \right\}.$

Since $B = \left\{ \begin{bmatrix} 2 \\ 3 \\ 1 \end{bmatrix}, \begin{bmatrix} 1 \\ -1 \\ 1 \end{bmatrix} \right\}$ is linear independent, B is a basis for W, so that $\dim(W) = 2$.

Exercise Set 3.4

If $B = \{\mathbf{v_1}, \ldots, \mathbf{v_n}\}$ is an ordered basis for a vector space, then for each vector $\mathbf{v}$ there are scalars $c_1, c_2, \ldots, c_n$ such that $\mathbf{v} = c_1\mathbf{v_1} + c_2\mathbf{v_2} + \cdots + c_n\mathbf{v_n}$. The unique scalars are called the coordinates of the vector relative to the ordered basis B, written as

$$[\mathbf{v}]_B = \begin{bmatrix} c_1 \\ c_2 \\ \vdots \\ c_n \end{bmatrix}.$$

If B is one of the standard bases of $\mathbb{R}^n$, then the coordinates of a vector are just the components of the vector. For example, since every vector in $\mathbb{R}^3$ can be written as

$$\mathbf{v} = \begin{bmatrix} x \\ y \\ z \end{bmatrix} = x\begin{bmatrix} 1 \\ 0 \\ 0 \end{bmatrix} + y\begin{bmatrix} 0 \\ 1 \\ 0 \end{bmatrix} + z\begin{bmatrix} 0 \\ 0 \\ 1 \end{bmatrix},$$

the coordinates relative to the standard basis are $[\mathbf{v}]_B = \begin{bmatrix} x \\ y \\ z \end{bmatrix}$. To find the coordinates relative to an ordered basis solve the usual vector equation

$$c_1\mathbf{v_1} + c_2\mathbf{v_2} + \cdots + c_n\mathbf{v_n} = \mathbf{v}$$

for the scalars $c_1, c_2, \ldots, c_n$. The order in which the vectors are given makes in difference when defining coordinates. For example, if $B = \left\{ \begin{bmatrix} 1 \\ 0 \end{bmatrix}, \begin{bmatrix} 0 \\ 1 \end{bmatrix} \right\}$ and $B' = \left\{ \begin{bmatrix} 0 \\ 1 \end{bmatrix}, \begin{bmatrix} 1 \\ 0 \end{bmatrix} \right\}$, then

$$[\mathbf{v}]_B = \left[\begin{bmatrix} x \\ y \end{bmatrix} \right]_B = \begin{bmatrix} x \\ y \end{bmatrix} \quad \text{and} \quad [\mathbf{v}]_{B'} = \left[\begin{bmatrix} x \\ y \end{bmatrix} \right]_{B'} = \begin{bmatrix} y \\ x \end{bmatrix}.$$

Given the coordinates relative to one basis $B = \{\mathbf{v_1}, \ldots, \mathbf{v_n}\}$ to find the coordinates of the same vector relative to a second basis $B' = \{\mathbf{v'_1}, \ldots, \mathbf{v'_n}\}$ a transition matrix can be used. To determine a transition matrix:

- Find the coordinates of each vector in B relative to the basis B'.

- Form the transition matrix where the column vectors are the coordinates found in the first step. That is,

$$[I]_B^{B'} = [\ [\mathbf{v_1}]_{B'} \ [\mathbf{v_2}]_{B'} \cdots [\mathbf{v_n}]_{B'} \ ].$$

- The coordinates of $\mathbf{v}$ relative to B' given the coordinates relative B are given by the formula

$$[\mathbf{v}]_{B'} = [I]_B^{B'} [\mathbf{v}]_B.$$

The transition matrix that changes coordinates from B' to B is given by

$$[I]_{B'}^B = ([I]_B^{B'})^{-1}.$$

Let $B = \left\{ \begin{bmatrix} 1 \\ 1 \end{bmatrix}, \begin{bmatrix} -1 \\ 1 \end{bmatrix} \right\}$ and $B' = \left\{ \begin{bmatrix} 1 \\ 2 \end{bmatrix}, \begin{bmatrix} -2 \\ -1 \end{bmatrix} \right\}$ be two bases for $\mathbb{R}^2$. The steps for finding the transition matrix from B to B' are:

- Since

$$\left[\begin{array}{cc|cc} 1 & -2 & 1 & -1 \\ 2 & -1 & 1 & 1 \end{array}\right] \xrightarrow{\text{reduces to}} \left[\begin{array}{cc|cc} 1 & -2 & 1 & -1 \\ 0 & 3 & -1 & 3 \end{array}\right],$$

so the coordinates of the two vectors in B relative to B' are $\left[\left[\begin{array}{c} 1 \\ 1 \end{array}\right]\right]_{B'} = \left[\begin{array}{c} 1/3 \\ -1/3 \end{array}\right]$ and $\left[\left[\begin{array}{c} -1 \\ 1 \end{array}\right]\right]_{B'} =$
$\left[\begin{array}{c} 1 \\ 1 \end{array}\right]$.

- $[I]_B^{B'} = \left[\begin{array}{cc} 1/3 & 1 \\ -1/3 & 1 \end{array}\right]$

- As an example, $\left[\left[\begin{array}{c} 3 \\ -2 \end{array}\right]\right]_{B'} = \left[\begin{array}{cc} 1/3 & 1 \\ -1/3 & 1 \end{array}\right]\left[\left[\begin{array}{c} 3 \\ -2 \end{array}\right]\right]_B = \left[\begin{array}{cc} 1/3 & 1 \\ -1/3 & 1 \end{array}\right]\left[\begin{array}{c} 1/2 \\ -5/2 \end{array}\right] = \left[\begin{array}{c} -7/3 \\ -8/3 \end{array}\right]$.

■ Solutions to Odd Exercises

1. The coordinates of $\left[\begin{array}{c} 8 \\ 0 \end{array}\right]$, relative to the basis B are the scalars c_1 and c_2 such that

$c_1\left[\begin{array}{c} 3 \\ 1 \end{array}\right] + c_2\left[\begin{array}{c} -2 \\ 2 \end{array}\right] = \left[\begin{array}{c} 8 \\ 0 \end{array}\right]$. The vector equation yields the linear system $\begin{cases} 3c_1 - 2c_2 = 8 \\ c_1 + 2c_2 = 0 \end{cases}$, which has the

unique solution $c_1 = 2$, and $c_2 = -1$. Hence, $[\mathbf{v}]_B = \left[\begin{array}{c} 2 \\ -1 \end{array}\right]$.

3. To find the coordinates we form and row reduce the matrix

$$\left[\begin{array}{ccc|c} 1 & 3 & 1 & 2 \\ -1 & -1 & 0 & -1 \\ 2 & 1 & 2 & 9 \end{array}\right] \longrightarrow \left[\begin{array}{ccc|c} 1 & 0 & 0 & 2 \\ 0 & 1 & 0 & -1 \\ 0 & 0 & 1 & 3 \end{array}\right], \text{ so that } [\mathbf{v}]_B = \left[\begin{array}{c} 2 \\ -1 \\ 3 \end{array}\right].$$

5. Since $c_1 + c_2(x-1) + c_3x^2 = 3 + 2x - 2x^2$ if and only if $c_1 - c_2 = 3, c_2 = 2$, and $c_3 = -2$, we have that
$[\mathbf{v}]_B = \left[\begin{array}{c} 5 \\ 2 \\ -2 \end{array}\right]$.

7. Since $c_1\left[\begin{array}{cc} 1 & -1 \\ 0 & 0 \end{array}\right] + c_2\left[\begin{array}{cc} 0 & 1 \\ 1 & 0 \end{array}\right] + c_3\left[\begin{array}{cc} 1 & 0 \\ 0 & -1 \end{array}\right] + c_4\left[\begin{array}{cc} 1 & 0 \\ -1 & 0 \end{array}\right] = \left[\begin{array}{cc} 1 & 3 \\ -2 & 2 \end{array}\right]$ if and only if

$\begin{cases} c_1 & + c_3 + c_4 & = 1 \\ -c_1 + c_2 & = 3 \\ c_2 & - c_4 & = -2 \\ -c_3 & = 2 \end{cases}$, and the linear system has the solution $c_1 = -1, c_2 = 2, c_3 = -2$, and $c_4 = 4$,

we have that $[\mathbf{v}]_B = \left[\begin{array}{c} -1 \\ 2 \\ -2 \\ 4 \end{array}\right]$.

9. $[\mathbf{v}]_{B_1} = \left[\begin{array}{c} -1/4 \\ 1/8 \end{array}\right]$; $[\mathbf{v}]_{B_2} = \left[\begin{array}{c} 1/2 \\ -1/2 \end{array}\right]$ $\qquad$ **11.** $[\mathbf{v}]_{B_1} = \left[\begin{array}{c} 1 \\ 2 \\ -1 \end{array}\right]$; $[\mathbf{v}]_{B_2} = \left[\begin{array}{c} 1 \\ 1 \\ 0 \end{array}\right]$

13. The column vectors for the transition matrix from a basis B_1 to a basis B_2 are the coordinate vectors
for the vectors in B_1 relative to B_2. Hence, $[I]_{B_1}^{B_2} = \left[\left[\begin{array}{c} 1 \\ 1 \end{array}\right]_{B_2} \left[\begin{array}{c} -1 \\ 1 \end{array}\right]_{B_2}\right] = \left[\begin{array}{cc} 1 & -1 \\ 1 & 1 \end{array}\right]$. Then this matrix

transforms coordinates relative to B_1 to coordinates relative to B_2, that is, $[\mathbf{v}]_{B_2} = [I]_{B_1}^{B_2}[\mathbf{v}]_{B_1} = \left[\begin{array}{c} -1 \\ 5 \end{array}\right]$.

15. $[I]_{B_1}^{B_2} = \begin{bmatrix} 3 & 2 & 1 \\ -1 & -2/3 & 0 \\ 0 & -1/3 & 0 \end{bmatrix}$; $[\mathbf{v}]_{B_2} =$

$[I]_{B_1}^{B_2}[\mathbf{v}]_{B_1} = \begin{bmatrix} -1 \\ 1 \\ 0 \end{bmatrix}$

17. Notice that the only difference in the bases B_1 and B_2 is the order in which the polynomials $1, x,$ and x^2 are given. As a result the column vectors of the transition matrix are the coordinate vectors only permuted. That is, $[I]_{B_1}^{B_2} = \begin{bmatrix} 0 & 0 & 1 \\ 1 & 0 & 0 \\ 0 & 1 & 0 \end{bmatrix}$.

Then $[\mathbf{v}]_{B_2} = [I]_{B_1}^{B_2}[\mathbf{v}]_{B_1} = \begin{bmatrix} 5 \\ 2 \\ 3 \end{bmatrix}$.

19. Since the equation $c_1 \begin{bmatrix} -1 \\ 1 \\ 1 \end{bmatrix} + c_2 \begin{bmatrix} 1 \\ 0 \\ 1 \end{bmatrix} + c_3 \begin{bmatrix} -1 \\ 1 \\ 0 \end{bmatrix} = \begin{bmatrix} a \\ b \\ c \end{bmatrix}$ gives

$$\begin{bmatrix} -1 & 1 & -1 & | & a \\ 1 & 0 & 1 & | & b \\ 1 & 1 & 0 & | & c \end{bmatrix} \longrightarrow \begin{bmatrix} 1 & 0 & 0 & | & -a-b+c \\ 0 & 1 & 0 & | & a+b \\ 0 & 0 & 1 & | & a+2b-c \end{bmatrix},$$ we have that $\begin{bmatrix} a \\ b \\ c \end{bmatrix}_B = \begin{bmatrix} -a-b+c \\ a+b \\ a+2b-c \end{bmatrix}$.

21. a. $[I]_{B_1}^{B_2} = \begin{bmatrix} 0 & 1 & 0 \\ 1 & 0 & 0 \\ 0 & 0 & 1 \end{bmatrix}$ **b.** $[\mathbf{v}]_{B_2} = [I]_{B_1}^{B_2} \begin{bmatrix} 1 \\ 2 \\ 3 \end{bmatrix} = \begin{bmatrix} 2 \\ 1 \\ 3 \end{bmatrix}$

23. a. $[I]_S^B = \begin{bmatrix} 1 & 1 \\ 0 & 2 \end{bmatrix}$ **b.** $\begin{bmatrix} 1 \\ 2 \end{bmatrix}_B = \begin{bmatrix} 3 \\ 4 \end{bmatrix}$; $\begin{bmatrix} 1 \\ 4 \end{bmatrix}_B = \begin{bmatrix} 5 \\ 8 \end{bmatrix}$; $\begin{bmatrix} 4 \\ 2 \end{bmatrix}_B = \begin{bmatrix} 6 \\ 4 \end{bmatrix}$;

$\begin{bmatrix} 4 \\ 4 \end{bmatrix}_B = \begin{bmatrix} 8 \\ 8 \end{bmatrix}$

c.

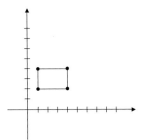

d.

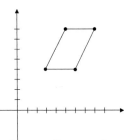

25. a. Since $\mathbf{u_1} = -\mathbf{v_1} + 2\mathbf{v_2}, \mathbf{u_2} = -\mathbf{v_1} + 2\mathbf{v_2} - \mathbf{v_3}$, and $\mathbf{u_3} = -\mathbf{v_2} + \mathbf{v_3}$, the coordinates of $\mathbf{u_1}, \mathbf{u_2},$ and $\mathbf{u_3}$ relative to B_2 are

$$[\mathbf{u_1}]_{B_2} = \begin{bmatrix} -1 \\ 2 \\ 0 \end{bmatrix}, [\mathbf{u_2}]_{B_2} = \begin{bmatrix} -1 \\ 2 \\ -1 \end{bmatrix}, [\mathbf{u_3}]_{B_2} = \begin{bmatrix} 0 \\ -1 \\ 1 \end{bmatrix}, \text{ so } [I]_{B_1}^{B_2} = \begin{bmatrix} -1 & -1 & 0 \\ 2 & 2 & -1 \\ 0 & -1 & 1 \end{bmatrix}.$$

b. $[2\mathbf{u_1} - 3\mathbf{u_2} + \mathbf{u_3}]_{B_2} = [I]_{B_1}^{B_2} \begin{bmatrix} 2 \\ -3 \\ 1 \end{bmatrix} = \begin{bmatrix} 1 \\ -3 \\ 4 \end{bmatrix}$

Exercise Set 3.5

1. a. Let $y = e^{rx}$, so that $y' = re^{rx}$ and $y'' = r^2 e^{rx}$. Substituting these into the differential equations gives the auxiliary equation $r^2 - 5r + 6 = 0$. Factoring, we have that $(r-3)(r-2) = 0$ and hence, two distinct solutions are $y_1 = e^{2x}$ and $y_2 = e^{3x}$.

b. Since $W[y_1, y_2](x) = \begin{vmatrix} e^{2x} & e^{3x} \\ 2e^{2x} & 3e^{3x} \end{vmatrix} = e^{5x} > 0$ for all x, the two solutions are linear independent. **c.** The general solution is the linear combination $y(x) = C_1 e^{2x} + C_2 e^{3x}$, where C_1 and C_2 are arbitrary constants.

3. a. Let $y = e^{rx}$, so that $y' = re^{rx}$ and $y'' = r^2 e^{rx}$. Substituting these into the differential equation gives the auxiliary equation $r^2 + 4r + 4 = 0$. Factoring, we have that $(r+2)^2 = 0$. Since the auxiliary equation has only one root of multiplicity 2, two distinct solutions are $y_1 = e^{-2x}$ and $y_2 = xe^{-2x}$.

b. Since $W[y_1, y_2](x) = \begin{vmatrix} e^{-2x} & xe^{-2x} \\ -2e^{-2x} & e^{-2x} - 2xe^{-2x} \end{vmatrix} = e^{-4x} > 0$ for all x, the two solutions are linearly independent. **c.** The general solution is the linear combination $y(x) = C_1 e^{-2x} + C_2 xe^{-2x}$, where C_1 and C_2 are arbitrary constants.

5. Let $y = e^{rx}$, so that $y' = re^{rx}$ and $y'' = r^2 e^{rx}$. Substituting these into the differential equation gives the auxiliary equation $r^2 - 2r + 1 = 0$. Factoring, we have that $(r-1)^2 = 0$. Since the auxiliary equation has only one root of multiplicity 2, two distinct and linearly independent solutions are $y_1 = e^x$ and $y_2 = xe^x$. The general solution is given by $y(x) = C_1 e^x + C_2 xe^x$. The initial value conditions now allow us to find the specific values for C_1 and C_2 to give the solution to the initial value problem. Specifically, since $y(0) = 1$, we have that $1 = y(x) = C_1 e^0 + C_2(0)e^0$, so $C_1 = 1$. Further, since $y'(x) = e^x + C_2(e^x + xe^x)$, and $y'(0) = 3$, we have that $3 = 1 + C_2$, so $C_2 = 2$. Then the solution to the initial value problem is $y(x) = e^x + 2xe^x$.

7. a. The auxiliary equation for $y'' - 4y' + 3y = 0$ is $r^2 - 4r + 3 = (r-3)(r-1) = 0$, so the complimentary solution is $y_c(x) = C_1 e^{3x} + C_2 e^x$.

b. Since $y_p'(x) = 2ax + b$ and $y_p''(x) = 2a$, we have that

$$2a - 4(2ax + b) + 3(ax^2 + bx + c) = 3x^2 + x + 2 \Leftrightarrow 3ax^2 + (3b - 8a)x + (2a + 3c - 4b) = 3x^2 + x + 2.$$

Equating coefficients of like terms, we have that $a = 1, b = 3$, and $c = 4$.

c. If $f(x) = y_c(x) + y_p(x)$, then

$$f'(x) = 3C_1 e^{3x} + C_2 e^x + 2x + 3 \quad \text{and} \quad f''(x) = 9C_1 e^{3x} + C_2 e^x + 2.$$

We then have that $f''(x) - 4f'(x) + 3f(x) = 3x^2 + x + 3$, so $y_c(x) + y_p(x)$ is a solution to the differential equation.

9. Since the damping coefficient is $c = 0$ and there is no external force acting on the system, so that $f(x) = 0$, the differential equation describing the problem has the form $my'' + ky = 0$. In addition, we have that $m = \frac{w}{g} = \frac{2}{32} = \frac{1}{16}$ and $k = \frac{w}{d} = \frac{2}{0.5} = 4$. Since the weight is pulled down 0.25 feet and then released the initial conditions on the system are $y(0) = 0.25$ and $y'(0) = 0$. The roots of the auxiliary equation $\frac{1}{16} y'' + 4y = 0$ are the complex values $r = \pm 8i$. Hence, the general solution is

$$y(x) = e^0 \left(C_1 \cos(8x) + C_2 \sin(8x) \right) = C_1 \cos(8x) + C_2 \sin(8x).$$

Applying the initial conditions we obtain $C_1 = 0.25$ and $C_2 = 0$. The equation of motion of the spring is $y(x) = \frac{1}{4} \cos(8x)$.

Review Exercises Chapter 3

1. Row reducing the matrix A with column vectors the given vectors gives

$$\begin{bmatrix} 1 & 0 & 0 & 2 \\ -2 & 1 & 0 & 3 \\ 0 & -1 & 1 & 4 \\ 2 & 3 & 4 & k \end{bmatrix} \longrightarrow \begin{bmatrix} 1 & 0 & 0 & 2 \\ 0 & 1 & 0 & 7 \\ 0 & 0 & 1 & 11 \\ 0 & 0 & 0 & k-69 \end{bmatrix}.$$

Hence, $\det(A) = k - 69$. Since the four vectors are linearly independent if and only if the determinant of A is nonzero, the vectors are a basis for $\mathbb{R}^4$ if and only if $k \neq 69$.

3. a. Since the sum of two 2×2 matrices and a scalar times a 2×2 matrix are 2×2 matrices, S is closed under vector addition and scalar multiplication. Hence, S is a subspace of $M_{2\times2}$.

b. Yes, let $a = 3, b = -2, c = 0$.

c. Since every matrix A in S can be written in the form,

$$A = a\begin{bmatrix} 1 & 1 \\ 0 & 1 \end{bmatrix} + b\begin{bmatrix} -1 & 0 \\ 1 & 0 \end{bmatrix} + c\begin{bmatrix} 0 & 0 \\ 1 & -1 \end{bmatrix}$$

and the matrices $\begin{bmatrix} 1 & 1 \\ 0 & 1 \end{bmatrix}, \begin{bmatrix} -1 & 0 \\ 1 & 0 \end{bmatrix}$, and $\begin{bmatrix} 0 & 0 \\ 1 & -1 \end{bmatrix}$ are linearly independent, a basis for S is $B = \left\{ \begin{bmatrix} 1 & 1 \\ 0 & 1 \end{bmatrix}, \begin{bmatrix} -1 & 0 \\ 1 & 0 \end{bmatrix}, \begin{bmatrix} 0 & 0 \\ 1 & -1 \end{bmatrix} \right\}$. **d.** The matrix $\begin{bmatrix} 0 & 1 \\ 2 & 1 \end{bmatrix}$ is not in S.

5. a. Consider the equation

$$c_1\mathbf{v_1} + c_2(\mathbf{v_1} + \mathbf{v_2}) + c_3(\mathbf{v_1} + \mathbf{v_2} + \mathbf{v_3}) = (c_1 + c_2 + c_3)\mathbf{v_1} + (c_2 + c_3)\mathbf{v_2} + c_3\mathbf{v_3} = \mathbf{0}.$$

Since S is linear independent, then $c_1 + c_2 + c_3 = 0, c_2 + c_3 = 0$, and $c_3 = 0$. The only solution to this system is the trivial solution, so that the set T is linearly independent. Since the set T consists of three linearly independent vectors in the three dimensional vector space V, the set T is a basis.

b. Consider the equation

$$c_1(-\mathbf{v_2} + \mathbf{v_3}) + c_2(3\mathbf{v_1} + 2\mathbf{v_2} + \mathbf{v_3}) + c_3(\mathbf{v_1} - \mathbf{v_2} + 2\mathbf{v_3}) = (3c_2 + c_3)\mathbf{v_1} + (-c_1 + 2c_2 - c_3)\mathbf{v_2} + (c_1 + c_2 + 2c_3)\mathbf{v_3} = \mathbf{0}.$$

Since S is linearly independent, we have that set W is linearly independent if and only if the linear system
$$\begin{cases} 3c_2 + c_3 &= 0 \\ -c_1 + 2c_2 - c_3 &= 0 \\ c_1 + c_2 + 2c_3 &= 0 \end{cases}$$ has only the trivial solution. Since $\begin{bmatrix} 0 & 3 & 1 \\ -1 & 2 & -1 \\ 1 & 1 & 2 \end{bmatrix} \longrightarrow \begin{bmatrix} -1 & 2 & -1 \\ 0 & 3 & 1 \\ 0 & 0 & 0 \end{bmatrix}$ the linear system has indefinitely many solutions and the set W is linearly dependent. Therefore, W is not basis.

7. Since $c \neq 0$ the vector $\mathbf{v_1}$ can be written as

$$\mathbf{v_1} = \left(\frac{-c_2}{c_1} \right)\mathbf{v_2} + \left(\frac{-c_3}{c_1} \right)\mathbf{v_3} + \cdots + \left(\frac{-c_n}{c_1} \right)\mathbf{v_n}.$$

Since $\mathbf{v}$ is a linear combination of the other vectors it does not contribute to the span of the set. Hence, $V = \mathbf{span}\{\mathbf{v_2}, \mathbf{v_3}, \ldots, \mathbf{v_n}\}$.

9. a. The set $B = \{\mathbf{u}, \mathbf{v}\}$ is a basis for $\mathbb{R}^2$ since it is linearly independent. To see this consider the equation $a\mathbf{u} + b\mathbf{v} = \mathbf{0}$. Now take the dot product of both sides first with $\mathbf{u}$ and then $\mathbf{v}$. That is,

$$\mathbf{u} \cdot (a\mathbf{u} + b\mathbf{v}) = 0 \Leftrightarrow a(\mathbf{u} \cdot \mathbf{u}) + b(\mathbf{u} \cdot \mathbf{v}) = 0 \Leftrightarrow a(u_1^2 + u_2^2) + b(\mathbf{u} \cdot \mathbf{v}) = 0.$$

Since $u_1^2 + u_2^2 = 1$ and $\mathbf{u} \cdot \mathbf{v} = 0$, we have that $a = 0$. Similarly, the equation $\mathbf{v} \cdot (a\mathbf{u} + b\mathbf{v}) = 0$ gives that $b = 0$. Hence, B is a set of two linearly independent vectors in $\mathbb{R}^2$ and therefore, is a basis.

b. If $[\mathbf{w}]_B = \begin{bmatrix} \alpha \\ \beta \end{bmatrix}$, then

$$\alpha\mathbf{u} + \beta\mathbf{v} = \begin{bmatrix} x \\ y \end{bmatrix} \Leftrightarrow \begin{bmatrix} \alpha u_1 + \beta v_1 \\ \alpha u_2 + \beta v_2 \end{bmatrix} = \begin{bmatrix} x \\ y \end{bmatrix}.$$

To solve the linear system for α and β by Cramer's Rule, we have that

$$\alpha = \frac{\begin{vmatrix} x & v_1 \\ y & v_2 \end{vmatrix}}{\begin{vmatrix} u_1 & v_1 \\ u_2 & v_2 \end{vmatrix}} = \frac{xv_2 - yv_1}{u_1v_2 - v_1u_2} \quad \text{and} \quad \beta = \frac{\begin{vmatrix} u_1 & x \\ u_2 & y \end{vmatrix}}{\begin{vmatrix} u_1 & v_1 \\ u_2 & v_2 \end{vmatrix}} = \frac{yu_1 - xu_2}{u_1v_2 - v_1u_2}.$$

Notice that $\begin{vmatrix} u_1 & v_1 \\ u_2 & v_2 \end{vmatrix} \neq 0$ since the vectors are linearly independent.

Chapter Test Chapter 3

1. F. Since

$$(c + d) \odot x = x + (c + d)$$

and

$$(c \odot x) \oplus (d \cdot x) = 2x + (c + d),$$

which do not always agree, V is not a vector space. Also $x \oplus y \neq y \oplus x$.

2. T

3. F. Only lines that pass through the origin are subspaces.

4. F. Since $\dim(M_{2\times2}) = 4$, every basis contains four matrices.

5. T

6. F. For example, the vector $\begin{bmatrix} -1 \\ -1 \end{bmatrix}$ is in S but $-\begin{bmatrix} -1 \\ -1 \end{bmatrix} = \begin{bmatrix} 1 \\ 1 \end{bmatrix}$ is not in S.

7. F. For example, the matrices $\begin{bmatrix} 1 & 0 \\ 0 & 0 \end{bmatrix}$ and $\begin{bmatrix} 0 & 0 \\ 0 & 1 \end{bmatrix}$ are in S but the sum $\begin{bmatrix} 1 & 0 \\ 0 & 1 \end{bmatrix}$ has determinant 1, so is not in S.

8. F. Since it is not possible to write x^3 as a linear combination of the given polynomials.

9. T

10. T

11. T

12. T

13. T

14. T

15. T

16. F. If a set spans a vector space, then adding more vectors can change whether the set is linearly independent or dependent but does not change the span.

17. F. A set of vectors with the number of vectors less than the dimension can be linearly independent but can not span the vector space. If the number of vectors exceeds the dimension, then the set is linearly dependent.

18. F. The intersection of two subspaces is always a subspace, but the union may not be a subspace.

19. T

20. T

21. T

22. T

23. T

24. T

25. T

26. F. $\begin{bmatrix} 1 \\ 1 \end{bmatrix}_{B_1} = \begin{bmatrix} 1 \\ 1/2 \end{bmatrix}$

27. T

28. T

29. T

30. F.

$$[x^3 + 2x^2 - x]_{B_1} = \begin{bmatrix} 0 \\ -1 \\ 2 \\ 1 \end{bmatrix}$$

31. T

32. F.

$$[x^3 + 2x^2 - x]_{B_2} = \begin{bmatrix} -1 \\ 2 \\ 0 \\ 1 \end{bmatrix}$$

33. T

34. F.

$$[(1+x)^2 - 3(x^2 + x - 1) + x^3]_{B_2}$$

$$= [4 - x - 2x^2 + x^3]_{B_2} = \begin{bmatrix} -1 \\ -2 \\ 4 \\ 1 \end{bmatrix}$$

35. T

4 Linear Transformations

Exercise Set 4.1

A linear transformation is a special kind of function (or mapping) defined from one vector space to another. To verify $T : V \longrightarrow W$ is a linear transformation from V to W, then we must show that T satisfies the two properties

$$T(\mathbf{u} + \mathbf{v}) = T(\mathbf{u}) + T(\mathbf{v}) \text{ and } T(c\mathbf{u}) = cT(\mathbf{u})$$

or equivalently just the one property

$$T(\mathbf{u} + c\mathbf{v}) = T(\mathbf{u}) + cT(\mathbf{v}).$$

The addition and scalar multiplication in $T(\mathbf{u} + c\mathbf{v})$ are the operations defined on V and in $T(\mathbf{u}) + cT(\mathbf{v})$ the operations defined on W. For example, $T : \mathbb{R}^2 \longrightarrow \mathbb{R}^2$ defined by

$$T\left(\begin{bmatrix} x \\ y \end{bmatrix} \right) = \begin{bmatrix} x + 2y \\ x - y \end{bmatrix} \quad \text{is a linear transformation.}$$

To see this compute the combination $\mathbf{u} + c\mathbf{v}$ of two vectors and then apply T. Notice that the definition of T requires the input of only one vector, so to apply T first simplify the expression. Then we need to consider

$$T\left(\begin{bmatrix} x_1 \\ y_1 \end{bmatrix} + c \begin{bmatrix} x_2 \\ y_2 \end{bmatrix} \right) = T\left(\begin{bmatrix} x_1 + cx_2 \\ y_1 + cy_2 \end{bmatrix} \right).$$

Next apply the definition of the mapping resulting in a vector with two components. To find the first component add the first component of the input vector to twice the second component and for the second component of the result subtract the components of the input vector. So

$$T\left(\begin{bmatrix} x_1 \\ y_1 \end{bmatrix} + c \begin{bmatrix} x_2 \\ y_2 \end{bmatrix} \right) = T\left(\begin{bmatrix} x_1 + cx_2 \\ y_1 + cy_2 \end{bmatrix} \right) = \begin{bmatrix} (x_1 + cx_2) + 2(y_1 + cy_2) \\ (x_1 + cx_2) - (y_1 + cy_2) \end{bmatrix}.$$

The next step is to rewrite the output vector in the correct form. This gives

$$\begin{bmatrix} (x_1 + cx_2) + 2(y_1 + cy_2) \\ (x_1 + cx_2) - (y_1 + cy_2) \end{bmatrix} = \begin{bmatrix} (x_1 + 2y_1) + c(x_2 + 2y_2) \\ (x_1 - y_1) + c(x_2 - y_2) \end{bmatrix}$$

$$= \begin{bmatrix} x_1 + 2y_1 \\ x_1 - y_1 \end{bmatrix} + c \begin{bmatrix} x_2 + 2y_2 \\ x_2 - y_2 \end{bmatrix} = T\left(\begin{bmatrix} x_1 \\ y_1 \end{bmatrix} \right) + cT\left(\begin{bmatrix} x_2 \\ y_2 \end{bmatrix} \right),$$

and hence T is a linear transformation. On the other hand a mapping defined by

$$T\left(\begin{bmatrix} x \\ y \end{bmatrix} \right) = \begin{bmatrix} x + 1 \\ y \end{bmatrix} \quad \text{is not a linear transformation}$$

since, for example,

$$T\left(\begin{bmatrix} x_1 \\ y_1 \end{bmatrix} + \begin{bmatrix} x_2 \\ y_2 \end{bmatrix} \right) = \begin{bmatrix} x_1 + x_2 + 1 \\ y_1 + y_2 \end{bmatrix}$$

$$\neq T\left(\begin{bmatrix} x_1 \\ y_1 \end{bmatrix} \right) + T\left(\begin{bmatrix} x_2 \\ y_2 \end{bmatrix} \right) = \begin{bmatrix} x_1 + 1 \\ y_1 \end{bmatrix} + \begin{bmatrix} x_2 + 1 \\ y_2 \end{bmatrix} = \begin{bmatrix} x_1 + x_2 + 2 \\ y_1 + y_2 \end{bmatrix}$$

for all pairs of vectors. Other useful observations made in Section 4.1 are:

- For every linear transformation $T(\mathbf{0}) = \mathbf{0}$.

- If A is an $m \times n$ matrix, then $T(\mathbf{v}) = A\mathbf{v}$ is a linear transformation from $\mathbb{R}^n$ to $\mathbb{R}^m$.

- $T(c_1\mathbf{v_1} + c_2\mathbf{v_2} + \cdots + c_n\mathbf{v_n}) = c_1T(\mathbf{v_1}) + c_2T(\mathbf{v_2}) + \cdots + c_nT(\mathbf{v_n})$

The third property can be used to find the image of a vector when the action of a linear transformation is known only on a specific set of vectors, for example on the vectors of a basis. For example, suppose that $T : \mathbb{R}^3 \longrightarrow \mathbb{R}^3$ is a linear transformation and

$$T\left(\begin{bmatrix} 1 \\ 1 \\ 1 \end{bmatrix}\right) = \begin{bmatrix} -1 \\ 2 \\ 0 \end{bmatrix}, T\left(\begin{bmatrix} 1 \\ 0 \\ 1 \end{bmatrix}\right) = \begin{bmatrix} 1 \\ 1 \\ 1 \end{bmatrix}, \text{ and } T\left(\begin{bmatrix} 0 \\ 1 \\ 1 \end{bmatrix}\right) = \begin{bmatrix} 2 \\ 3 \\ -1 \end{bmatrix}.$$

Then the image of an arbitrary input vector can be found since $\left\{ \begin{bmatrix} 1 \\ 1 \\ 1 \end{bmatrix}, \begin{bmatrix} 1 \\ 0 \\ 1 \end{bmatrix}, \begin{bmatrix} 0 \\ 1 \\ 1 \end{bmatrix} \right\}$ is a basis for $\mathbb{R}^3$.

For example, let's find the image of the vector $\begin{bmatrix} 1 \\ -2 \\ 0 \end{bmatrix}$. The first step is to write the input vector in terms of the basis vectors, so

$$\begin{bmatrix} 1 \\ -2 \\ 0 \end{bmatrix} = -\begin{bmatrix} 1 \\ 1 \\ 1 \end{bmatrix} + 2\begin{bmatrix} 1 \\ 0 \\ 1 \end{bmatrix} - \begin{bmatrix} 0 \\ 1 \\ 1 \end{bmatrix}.$$

Then use the linearity properties of T to obtain

$$T\left(\begin{bmatrix} 1 \\ -2 \\ 0 \end{bmatrix}\right) = T\left(-\begin{bmatrix} 1 \\ 1 \\ 1 \end{bmatrix} + 2\begin{bmatrix} 1 \\ 0 \\ 1 \end{bmatrix} - \begin{bmatrix} 0 \\ 1 \\ 1 \end{bmatrix}\right) = -T\left(\begin{bmatrix} 1 \\ 1 \\ 1 \end{bmatrix}\right) + 2T\left(\begin{bmatrix} 1 \\ 0 \\ 1 \end{bmatrix}\right) - T\left(\begin{bmatrix} 0 \\ 1 \\ 1 \end{bmatrix}\right)$$

$$= -\begin{bmatrix} -1 \\ 2 \\ 0 \end{bmatrix} + 2\begin{bmatrix} 1 \\ 1 \\ 1 \end{bmatrix} - \begin{bmatrix} 2 \\ 3 \\ -1 \end{bmatrix} = \begin{bmatrix} 1 \\ -3 \\ 3 \end{bmatrix}.$$

■ Solutions to Odd Exercises

1. Let $\mathbf{u} = \begin{bmatrix} u_1 \\ u_2 \end{bmatrix}$ and $\mathbf{v} = \begin{bmatrix} v_1 \\ v_2 \end{bmatrix}$ be vectors in $\mathbb{R}^2$ and c a scalar. Since

$$T(\mathbf{u} + c\mathbf{v}) = T\left(\begin{bmatrix} u_1 + cv_1 \\ u_2 + cv_2 \end{bmatrix}\right) = \begin{bmatrix} u_2 + cv_2 \\ u_1 + cv_1 \end{bmatrix} = \begin{bmatrix} u_2 \\ u_1 \end{bmatrix} + c\begin{bmatrix} v_2 \\ v_1 \end{bmatrix} = T(\mathbf{u}) + cT(\mathbf{v}),$$

then T is a linear transformation.

3. Let $\mathbf{u} = \begin{bmatrix} u_1 \\ u_2 \end{bmatrix}$ and $\mathbf{v} = \begin{bmatrix} v_1 \\ v_2 \end{bmatrix}$ be vectors in $\mathbb{R}^2$. Since

$$T(\mathbf{u} + \mathbf{v}) = T\left(\begin{bmatrix} u_1 + v_1 \\ u_2 + v_2 \end{bmatrix}\right) = \begin{bmatrix} u_1 + v_1 \\ u_1^2 + 2u_2v_2 + v_2^2 \end{bmatrix} \text{ and } \begin{bmatrix} u_1 + v_1 \\ u_2^2 + v_2^2 \end{bmatrix} = T(\mathbf{u}) + T(\mathbf{v}),$$

which do not agree for all vectors, T is not a linear transformation.

5. Since $T\left(\begin{bmatrix} u_1 \\ u_2 \end{bmatrix} + c\begin{bmatrix} v_1 \\ v_2 \end{bmatrix}\right) = \begin{bmatrix} u_1 + cv_1 \\ 0 \end{bmatrix} = T\left(\begin{bmatrix} u_1 \\ u_2 \end{bmatrix}\right) + cT\left(\begin{bmatrix} v_1 \\ v_2 \end{bmatrix}\right)$, for all pairs of vectors and scalars c, T is a linear transformation.

7. Since $T(x + y) = T(x) + T(y)$, if and only if at least one of x or y is zero, T is not a linear transformation.

9. Since $T\left(c\begin{bmatrix} x \\ y \end{bmatrix}\right) = c^2(x^2 + y^2) =$ $cT\left(\begin{bmatrix} x \\ y \end{bmatrix}\right) = c(x^2+y^2)$ if and only if $c = 1$ or $\begin{bmatrix} x \\ y \end{bmatrix} = \begin{bmatrix} 0 \\ 0 \end{bmatrix}$, T is not a linear transformation.

11. Since $T(\mathbf{0}) \neq \mathbf{0}$, T is not a linear transformation.

13. Since

$$T(p(x) + q(x)) = 2(p''(x) + q''(x)) - 3(p'(x) + q'(x)) + (p(x) + q(x))$$
$$= (2p''(x) - 3p'(x) + p(x)) + (2q''(x) - 3q'(x) + q(x)) = T(p(x)) + T(q(x))$$

and similarly, $T(cp(x)) = cT(p(x))$ for all scalars c, T is a linear transformation.

15. If A is a 2×2 matrix, then $\det(cA) = c^2\det(A)$, then for example, if $c = 2$ and $\det(A) \neq 0$, we have that $T(2A) = 4T(A) \neq 2T(A)$ and hence, T is not a linear transformation.

17. a. $T(\mathbf{u}) = \begin{bmatrix} 2 \\ 3 \end{bmatrix}$; $T(\mathbf{v}) = \begin{bmatrix} -2 \\ -2 \end{bmatrix}$ **b.** $T(\mathbf{u} + \mathbf{v}) = T\left(\begin{bmatrix} 0 \\ 1 \end{bmatrix}\right) = \begin{bmatrix} 0 \\ 1 \end{bmatrix} = T(\mathbf{u}) + T(\mathbf{v})$
c. Yes, T is a linear transformation.

19. a. $T(\mathbf{u}) = \begin{bmatrix} 0 \\ 0 \end{bmatrix}$; $T(\mathbf{v}) = \begin{bmatrix} 0 \\ -1 \end{bmatrix}$ **b.** $T(\mathbf{u} + \mathbf{v}) = \begin{bmatrix} -1 \\ -1 \end{bmatrix} \neq T(\mathbf{u}) + T(\mathbf{v}) = \begin{bmatrix} 0 \\ -1 \end{bmatrix}$
c. By part (b), T is not a linear transformation.

21. Since $\begin{bmatrix} 1 \\ -3 \end{bmatrix} = \begin{bmatrix} 1 \\ 0 \end{bmatrix} - 3\begin{bmatrix} 0 \\ 1 \end{bmatrix}$ and T is a linear transformation, we have that

$$T\left(\begin{bmatrix} 1 \\ -3 \end{bmatrix}\right) = T\left(\begin{bmatrix} 1 \\ 0 \end{bmatrix} - 3\begin{bmatrix} 0 \\ 1 \end{bmatrix}\right) = T\left(\begin{bmatrix} 1 \\ 0 \end{bmatrix}\right) - 3T\left(\begin{bmatrix} 0 \\ 1 \end{bmatrix}\right) = \begin{bmatrix} 5 \\ -9 \end{bmatrix}.$$

23. Since $T(-3 + x - x^2) = T(-3(1) + 1(x) + (-1)x^2)$ and T is a linear operator, then $T(-3 + x - x^2) = -3(1 + x) + (2 + x^2) - (x - 3x^2) = -1 - 4x + 4x^2$.

25. Since $\left\{\begin{bmatrix} 1 \\ 1 \end{bmatrix}, \begin{bmatrix} -1 \\ 0 \end{bmatrix}\right\}$ is a basis for $\mathbb{R}^2$ and T is a linear operator, then it is possible to find $T(\mathbf{v})$ for every vector in $\mathbb{R}^2$. In particular, $T\left(\begin{bmatrix} 3 \\ 7 \end{bmatrix}\right) = T\left(7\begin{bmatrix} 1 \\ 1 \end{bmatrix} + 4\begin{bmatrix} -1 \\ 0 \end{bmatrix}\right) = \begin{bmatrix} 22 \\ -11 \end{bmatrix}$.

27. a. Since the polynomial $2x^2 - 3x + 2$ cannot be written as a linear combination of x^2, $-3x$, and $-x^2 + 3x$, from the given information the value of $T(2x^2 - 3x + 2)$ can not be determined. That is, the equation $c_1x^2 + c_2(-3x) + c_3(-x^2 + 3x) = 2x^2 - 2x + 1$ is equivalent to $(c_1 - c_3)x^2 + (-3c_2 + 3c_3)x = 2x^2 - 3x + 2$, which is not possible. **b.** $T(3x^2 - 4x) = T\left(3x^2 + \frac{4}{3}(-3x)\right) = 3T(x^2) + \frac{4}{3}T(-3x) = \frac{4}{3}x^2 + 6x - \frac{13}{3}$.

29. a. If $A = \begin{bmatrix} -1 & 0 \\ 0 & -1 \end{bmatrix}$, then $T\left(\begin{bmatrix} x \\ y \end{bmatrix}\right) = A\begin{bmatrix} x \\ y \end{bmatrix} = \begin{bmatrix} -x \\ -y \end{bmatrix}$. **b.** $T(\mathbf{e_1}) = \begin{bmatrix} -1 \\ 0 \end{bmatrix}$ and $T(\mathbf{e_2}) = \begin{bmatrix} 0 \\ -1 \end{bmatrix}$. Observe that these are the column vectors of A.

31. A vector $\begin{bmatrix} x \\ y \\ z \end{bmatrix}$ in $\mathbb{R}^3$ is mapped to the zero vector if and only if $\begin{bmatrix} x + y \\ x - y \end{bmatrix} = \begin{bmatrix} 0 \\ 0 \end{bmatrix}$, that is, if and only if $x = y = 0$. Consequently, $T\left(\begin{bmatrix} 0 \\ 0 \\ z \end{bmatrix}\right) = \begin{bmatrix} 0 \\ 0 \end{bmatrix}$, for all $z \in \mathbb{R}$.

33. a. Since

$$\left[\begin{array}{rrr|r} 1 & -1 & 2 & 0 \\ 2 & 3 & -1 & 0 \\ -1 & 2 & -2 & 0 \end{array}\right] \longrightarrow \left[\begin{array}{rrr|r} 1 & -1 & 2 & 0 \\ 0 & 5 & -5 & 0 \\ 0 & 0 & 1 & 0 \end{array}\right],$$

the zero vector is the only vector in $\mathbb{R}^3$ such that $T\left(\begin{bmatrix} x \\ y \\ z \end{bmatrix}\right) = \begin{bmatrix} 0 \\ 0 \end{bmatrix}$.

b. Since

$$\left[\begin{array}{rrr|r} 1 & -1 & 2 & 7 \\ 2 & 3 & -1 & -6 \\ -1 & 2 & -2 & -9 \end{array}\right] \longrightarrow \left[\begin{array}{rrr|r} 1 & 0 & 0 & 1 \\ 0 & 1 & 0 & -2 \\ 0 & 0 & 1 & 2 \end{array}\right], \text{ then } T\left(\begin{bmatrix} 1 \\ -2 \\ 2 \end{bmatrix}\right) = \begin{bmatrix} 7 \\ -6 \\ -9 \end{bmatrix}.$$

35. Since $T(c\mathbf{v} + \mathbf{w}) = \begin{bmatrix} cT_1(\mathbf{v}) + T_1(\mathbf{w}) \\ cT_2(\mathbf{v}) + T_2(\mathbf{w}) \end{bmatrix} = c\begin{bmatrix} T_1(\mathbf{v}) \\ T_2(\mathbf{v}) \end{bmatrix} + \begin{bmatrix} T_1(\mathbf{w}) \\ T_2(\mathbf{w}) \end{bmatrix} = cT(\mathbf{v}) + T(\mathbf{w})$, then T is a linear transformation.

37. Since $T(kA + C) = (kA + C)B - B(kA + C) = kAB - kBA + CB - BC = kT(A) + T(C)$, then T is a linear operator.

39. a. Using the properties of the Riemann Integral, we have that

$$T(cf + g) = \int_0^1 (cf(x) + g(x))\, dx = \int_0^1 cf(x)dx + \int_0^1 g(x)dx = c\int_0^1 f(x)dx + \int_0^1 g(x)dx = cT(f) + T(g)$$

so T is a linear operator. **b.** $T(2x^2 - x + 3) = \frac{19}{6}$

41. Since $\{\mathbf{v}, \mathbf{w}\}$ is linear independent $\mathbf{v} \neq \mathbf{0}$ and $\mathbf{w} \neq \mathbf{0}$. Hence, if either $T(\mathbf{v}) = 0$ or $T(\mathbf{w}) = 0$, then the conclusion holds. Now assume that $T(\mathbf{v})$ and $T(\mathbf{w})$ are linearly dependent and not zero. So there exist scalars a and b, not both 0, such that $aT(\mathbf{v}) + bT(\mathbf{w}) = \mathbf{0}$. Since $\mathbf{v}$ and $\mathbf{w}$ are linearly independent, then $a\mathbf{v} + b\mathbf{w} \neq \mathbf{0}$. Hence, since T is linear, then $aT(\mathbf{v}) + bT(\mathbf{w}) = T(a\mathbf{v} + b\mathbf{w}) = \mathbf{0}$, and we have shown that $T(\mathbf{u}) = \mathbf{0}$ has a nontrivial solution.

43. Let $T(\mathbf{v}) = \mathbf{0}$ for all $\mathbf{v}$ in $\mathbb{R}^3$.

Exercise Set 4.2

If $T : V \longrightarrow W$ is a linear transformation, then the null space is the subspace of all vectors in V that are mapped to the zero vector in W and the range of T is the subspace of W consisting of all images of vectors from V. Any transformation defined by a matrix product is a linear transformation. For example, $T : \mathbb{R}^3 \longrightarrow \mathbb{R}^3$ defined by

$$T\left(\begin{bmatrix} x_1 \\ x_2 \\ x_3 \end{bmatrix}\right) = A\begin{bmatrix} x_1 \\ x_2 \\ x_3 \end{bmatrix} = \begin{bmatrix} 1 & 3 & 0 \\ 2 & 0 & 3 \\ 2 & 0 & 3 \end{bmatrix}\begin{bmatrix} x_1 \\ x_2 \\ x_3 \end{bmatrix}$$

is a linear transformation. The null space of T, denoted by $N(T)$, is the null space of the matrix, $N(A) = \{\mathbf{x} \in \mathbb{R}^3 \mid A\mathbf{x} = \mathbf{0}\}$. Since

$$T\left(\begin{bmatrix} x_1 \\ x_2 \\ x_3 \end{bmatrix}\right) = \begin{bmatrix} 1 & 3 & 0 \\ 2 & 0 & 3 \\ 2 & 0 & 3 \end{bmatrix}\begin{bmatrix} x_1 \\ x_2 \\ x_3 \end{bmatrix} = x_1\begin{bmatrix} 1 \\ 2 \\ 2 \end{bmatrix} + x_2\begin{bmatrix} 3 \\ 0 \\ 0 \end{bmatrix} + x_3\begin{bmatrix} 0 \\ 3 \\ 3 \end{bmatrix},$$

the range of T, denoted by $R(T)$ is the column space of A, $\mathbf{col}(A)$. Since

$$\begin{bmatrix} 1 & 3 & 0 \\ 2 & 0 & 3 \\ 2 & 0 & 3 \end{bmatrix} \xrightarrow{\text{reduces to}} \begin{bmatrix} 1 & 3 & 0 \\ 0 & -6 & 3 \\ 0 & 0 & 0 \end{bmatrix}$$

the homogeneous equation $A\mathbf{x} = \mathbf{0}$ has infinitely many solutions given by $x_1 = -\frac{3}{2}x_3, x_2 = \frac{1}{2}x_3$, and x_3 a free variable. So the null space is $\left\{ t \begin{bmatrix} -3/2 \\ 1/2 \\ 1 \end{bmatrix} \middle| \; t \in \mathbb{R} \right\}$, which is a line that passes through the origin in three space. Also since the pivots in the reduced matrix are in columns one and two, a basis for the range is $\left\{ \begin{bmatrix} 1 \\ 2 \\ 2 \end{bmatrix}, \begin{bmatrix} 3 \\ 0 \\ 0 \end{bmatrix} \right\}$ and hence, the range is a plane in three space. Notice that in this example, $3 = \dim(\mathbb{R}^3) = \dim(R(T)) + \dim(N(T))$. This is a fundamental theorem that if $T : V \longrightarrow W$ is a linear transformation defined on finite dimensional vector spaces, then

$$\dim(V) = \dim(R(T)) + \dim(N(T)).$$

If the mapping is given as a matrix product $T(\mathbf{v}) = A\mathbf{v}$ such that A is a $m \times n$ matrix, then this result is written as

$$n = \mathbf{rank}(A) + \mathbf{nullity}(A).$$

A number of useful statements are added to the list of equivalences concerning $n \times n$ linear systems:

A is invertible $\Leftrightarrow A\mathbf{x} = \mathbf{b}$ has a unique solution for every $\mathbf{b} \Leftrightarrow A\mathbf{x} = \mathbf{0}$ has only the trivial solution
$\qquad \Leftrightarrow A$ is row equivalent to $I \Leftrightarrow \det(A) \neq 0 \Leftrightarrow$ the column vectors of A are linearly independent
$\qquad \Leftrightarrow$ the column vectors of A span $\mathbb{R}^n \Leftrightarrow$ the column vectors of A are a basis for $\mathbb{R}^n$
$\qquad \Leftrightarrow \mathbf{rank}(A) = n \Leftrightarrow R(A) = \mathbf{col}(A) = \mathbb{R}^n \Leftrightarrow N(A) = \{\mathbf{0}\} \Leftrightarrow \mathbf{row}(A) = \mathbb{R}^n$
$\qquad \Leftrightarrow$ the number of pivot columns in the row echelon form of A is n.

■ Solutions to Odd Exercises

1. Since $T(\mathbf{v}) = \begin{bmatrix} 0 \\ 0 \end{bmatrix}$, $\mathbf{v}$ is in $N(T)$.

3. Since $T(\mathbf{v}) = \begin{bmatrix} -5 \\ 10 \end{bmatrix}$, $\mathbf{v}$ is not in $N(T)$.

5. Since $p'(x) = 2x - 3$ and $p''(x) = 2$, then $T(p(x)) = 2x$, so $p(x)$ is not in $N(T)$.

7. Since $T(p(x)) = -2x$, $p(x)$ is not in $N(T)$.

9. Since $\begin{bmatrix} 1 & 0 & 2 & | & 1 \\ 2 & 1 & 3 & | & 3 \\ 1 & -1 & 3 & | & 0 \end{bmatrix} \longrightarrow \begin{bmatrix} 1 & 0 & 2 & | & 1 \\ 0 & 1 & -1 & | & 1 \\ 0 & 0 & 0 & | & 0 \end{bmatrix}$ there are infinitely many vectors that are mapped to $\begin{bmatrix} 1 \\ 3 \\ 0 \end{bmatrix}$. For example, $T\left(\begin{bmatrix} -1 \\ 2 \\ 1 \end{bmatrix} \right) = \begin{bmatrix} 1 \\ 3 \\ 0 \end{bmatrix}$ and hence, $\begin{bmatrix} 1 \\ 3 \\ 0 \end{bmatrix}$ is in $R(T)$.

11. Since $\begin{bmatrix} 1 & 0 & 2 & | & -1 \\ 2 & 1 & 3 & | & 1 \\ 1 & -1 & 3 & | & -2 \end{bmatrix} \longrightarrow \begin{bmatrix} 1 & 0 & 2 & | & 0 \\ 0 & 1 & -1 & | & 0 \\ 0 & 0 & 0 & | & 1 \end{bmatrix}$, the linear system is inconsistent, so the vector $\begin{bmatrix} -1 \\ 1 \\ -2 \end{bmatrix}$ is not in $R(T)$.

13. Since $T\left(\begin{bmatrix} 1 & 0 \\ -2 & -1 \end{bmatrix} \right) = A$, the matrix A is in $R(T)$.

15. The matrix A is not in $R(T)$.

17. A vector $\mathbf{v} = \begin{bmatrix} x \\ y \end{bmatrix}$ is in the null space, if and only if $3x + y = 0$ and $y = 0$. That is, $N(T) = \left\{ \begin{bmatrix} 0 \\ 0 \end{bmatrix} \right\}$. Hence, the null space has dimension 0, so does not have a basis.

19. Since $\begin{bmatrix} x+2z \\ 2x+y+3z \\ x-y+3z \end{bmatrix} = \begin{bmatrix} 0 \\ 0 \\ 0 \end{bmatrix}$ if and only if $x = -2z$ and $y = z$ every vector in the null space has the

form $\begin{bmatrix} -2z \\ z \\ z \end{bmatrix}$. Hence, a basis for the null space is $\left\{ \begin{bmatrix} -2 \\ 1 \\ 1 \end{bmatrix} \right\}$.

21. Since $N(T) = \left\{ \begin{bmatrix} 2s+t \\ s \\ t \end{bmatrix} \middle| s,t \in \mathbb{R} \right\}$, a basis for the null space is $\left\{ \begin{bmatrix} 2 \\ 1 \\ 0 \end{bmatrix}, \begin{bmatrix} 1 \\ 0 \\ 1 \end{bmatrix} \right\}$.

23. Since $T(p(x)) = 0$ if and only if $p(0) = 0$ a polynomial is in the null space if and only if it has the form $ax^2 + bx$. A basis for the null space is $\{x, x^2\}$.

25. Since $\det\left(\begin{bmatrix} 1 & 1 & 2 \\ 0 & 1 & -1 \\ 2 & 0 & 1 \end{bmatrix} \right) = -5$, the column vectors of the matrix are a basis for the column space of the matrix. Since the column space of the matrix is $R(T)$, then a basis for the range of T is

$\left\{ \begin{bmatrix} 1 \\ 0 \\ 2 \end{bmatrix}, \begin{bmatrix} 1 \\ 1 \\ 0 \end{bmatrix}, \begin{bmatrix} 2 \\ -1 \\ 1 \end{bmatrix} \right\}$.

27. Since the range of T is the xy-plane in $\mathbb{R}^3$, a basis for the range is $\left\{ \begin{bmatrix} 1 \\ 0 \\ 0 \end{bmatrix}, \begin{bmatrix} 0 \\ 1 \\ 0 \end{bmatrix} \right\}$.

29. Since $R(T) = \mathcal{P}_2$, then a basis for the range is $\{1, x, x^2\}$.

31. a. The vector $\mathbf{w}$ is in the range of T if the linear system

$$c_1 \begin{bmatrix} -2 \\ 1 \\ 1 \end{bmatrix} + c_2 \begin{bmatrix} 0 \\ 1 \\ -1 \end{bmatrix} + c_3 \begin{bmatrix} -2 \\ 2 \\ 0 \end{bmatrix} = \begin{bmatrix} -6 \\ 5 \\ 0 \end{bmatrix}$$

has a solution. But $\begin{bmatrix} -2 & 0 & -2 & | & -6 \\ 1 & 1 & 2 & | & 5 \\ 1 & -1 & 0 & | & 0 \end{bmatrix} \longrightarrow \begin{bmatrix} -2 & 0 & -2 & | & -6 \\ 0 & 1 & 1 & | & 2 \\ 0 & 0 & 0 & | & -1 \end{bmatrix}$, so that the linear system is inconsis-

tent. Hence, $\begin{bmatrix} -6 \\ 5 \\ 0 \end{bmatrix}$ is not in $R(t)$.

b. Since $\begin{vmatrix} -2 & 0 & -2 \\ 1 & 1 & 2 \\ 1 & -1 & 0 \end{vmatrix} = 0$, the column vectors are linearly dependent. To trim the vectors to a basis for the range, we have that

$$\begin{bmatrix} -2 & 0 & -2 \\ 1 & 1 & 2 \\ 1 & -1 & 0 \end{bmatrix} \longrightarrow \begin{bmatrix} -2 & 0 & -2 \\ 0 & 1 & 1 \\ 0 & 0 & 0 \end{bmatrix}.$$

Since the pivots are in columns one and two, a basis for the range is $\left\{ \begin{bmatrix} -2 \\ 1 \\ 1 \end{bmatrix}, \begin{bmatrix} 0 \\ 1 \\ -1 \end{bmatrix} \right\}$.

c. Since $\dim(N(T)) + \dim(R(T)) = \dim(\mathbb{R}^3) = 3$ and $\dim(R(T)) = 2$, then $\dim(N(T)) = 1$.

33. a. The polynomial $2x^2 - 4x + 6$ is not in $R(T)$. **b.** A basis for the range is $\{T(x), T(x^2)\} = \{-2x + 1, x^2 + x\}$.

35. Any linear transformations that maps three space to the entire xy-plane will work. For example, the

mapping to the xy-plane is $T\left(\begin{bmatrix} x \\ y \\ z \end{bmatrix} \right) = \begin{bmatrix} x \\ y \end{bmatrix}$.

37. a. The range $R(T)$ is the subspace of $\mathcal{P}_n$ consisting of all polynomials of degree $n-1$ or less. **b.** $\dim(R(T)) = n$ **c.** Since $\dim(R(T)) + \dim(N(T)) = \dim(\mathcal{P}_n) = n+1$, then $\dim(N(T)) = 1$.

39. a. $\dim(R(T)) = 2$ **b.** $\dim(N(T)) = 1$

41. If $B = \begin{bmatrix} a & b \\ c & d \end{bmatrix}$, then $T(B) = AB - BA = \begin{bmatrix} 0 & 2b \\ -2c & 0 \end{bmatrix}$, so that

$$N(T) = \left\{ \begin{bmatrix} a & 0 \\ 0 & d \end{bmatrix} \middle| a, d \in \mathbb{R} \right\} = \left\{ a \begin{bmatrix} 1 & 0 \\ 0 & 0 \end{bmatrix} + d \begin{bmatrix} 0 & 0 \\ 0 & 1 \end{bmatrix} \middle| a, d \in \mathbb{R} \right\}.$$ Hence a basis for $N(T)$ is

$$\left\{ \begin{bmatrix} 1 & 0 \\ 0 & 0 \end{bmatrix}, \begin{bmatrix} 0 & 0 \\ 0 & 1 \end{bmatrix} \right\}.$$

43. a. Notice that $(A + A^t)^t = A^t + A = A + A^t$, so that the range of T is a subset of the symmetric matrices. Also if B is any symmetric matrix, then $T\left(\frac{1}{2}B\right) = \frac{1}{2}B + \frac{1}{2}B^t = B$. Therefore, $R(T)$ is the set of all symmetric matrices. **b.** Since a matrix A is in $N(T)$ if and only if $T(A) = A + A^t = \mathbf{0}$, which is if and only if $A = -A^t$, then the null space of T is the set of skew-symmetric matrices.

45. If the matrix A is invertible and B is any $n \times n$ matrix, then $T(A^{-1}B) = A(A^{-1}B) = B$, so $R(T) = M_{n \times n}$.

Exercise Set 4.3

An isomorphism between vector spaces establishes a one-to-one correspondence between the vector spaces. If $T : V \longrightarrow W$ is a one-to-one and onto linear transformation, then T is called an isomorphism. A mapping is one-to-one if and only if $N(T) = \{\mathbf{0}\}$ and is onto if and only if $R(T) = W$. If $\{\mathbf{v_1}, \ldots, \mathbf{v_n}\}$ is a basis for V and $T : V \longrightarrow W$ is a linear transformation, then $R(T) = \mathbf{span}\{T(\mathbf{v_1}), \ldots, T(\mathbf{v_n})\}$. If in addition, T is one-to-one, then $\{T(\mathbf{v_1}), \ldots, T(\mathbf{v_n})\}$ is a basis for $R(T)$. The main results of Section 4.3 are:

- If V is a vector space with $\dim(V) = n$, then V is isomorphic to $\mathbb{R}^n$.

- If V and W are vector spaces of dimension n, then V and W are isomorphic.

For example, there is a correspondence between the very different vector spaces $\mathcal{P}_3$ and $M_{2 \times 2}$. To define the isomorphism, start with the standard basis $S = \{1, x, x^2, x^3\}$ for $\mathcal{P}_3$. Since every polynomial $a + bx + cx^2 + dx^3 = a(1) + b(x) + c(x^2) + d(x^3)$ use the coordinate map

$$a + bx + cx^2 + dx^3 \xrightarrow{L_1} [a + bx + cx^2 + dx^3]_S = \begin{bmatrix} a \\ b \\ c \\ d \end{bmatrix} \text{ followed by } \begin{bmatrix} a \\ b \\ c \\ d \end{bmatrix} \xrightarrow{L_2} \begin{bmatrix} a & b \\ c & d \end{bmatrix},$$

so that the composition $L_2(L_1(a + bx + cx^2 + dx^3)) = \begin{bmatrix} a & b \\ c & d \end{bmatrix}$ defines an isomorphism between $\mathcal{P}_3$ and $M_{2 \times 2}$.

■ Solutions to Odd Exercises

1. Since $N(T) = \left\{ \begin{bmatrix} 0 \\ 0 \end{bmatrix} \right\}$, then T is one-to-one.   **3.** Since $N(T) = \left\{ \begin{bmatrix} 0 \\ 0 \\ 0 \end{bmatrix} \right\}$, then T is one-to-one.

5. Let $p(x) = ax^2 + bx + c$, so that $p'(x) = 2ax + b$. Then

$$T(p(x)) = 2ax + b - ax^2 - bx - c = -ax^2 + (2a - b)x + (b - c) = 0$$

if and only if $a = 0, 2a - b = 0, b - c = 0$. That is, $p(x)$ is in $N(T)$ if and only if $p(x) = 0$. Hence, T is one-to-one.

7. A vector $\begin{bmatrix} a \\ b \end{bmatrix}$ is in the range of T if the linear system $\begin{cases} 3x - y = a \\ x + y = b \end{cases}$ has a solution. Since the linear system is consistent for every vector $\begin{bmatrix} a \\ b \end{bmatrix}$, T is onto $\mathbb{R}^2$. Notice the result also follows from $\det\left(\begin{bmatrix} 3 & -1 \\ 1 & 1 \end{bmatrix}\right) = 4$, so the inverse exists.

9. The linear operator T is onto $\mathbb{R}^3$.

11. Since $T(\mathbf{e_1}) = \begin{bmatrix} -1 \\ 3 \end{bmatrix}$ and $T(\mathbf{e_2}) = \begin{bmatrix} -2 \\ 0 \end{bmatrix}$ are two linear independent vectors in $\mathbb{R}^2$, they form a basis.

13. Since $T(\mathbf{e_1}) = \begin{bmatrix} 3 \\ -3 \end{bmatrix}$ and $T(\mathbf{e_2}) = \begin{bmatrix} -1 \\ -1 \end{bmatrix}$ are two linear independent vectors in $\mathbb{R}^2$, they form a basis.

15. Since $T(\mathbf{e_1}) = \begin{bmatrix} -1 \\ 0 \\ 0 \end{bmatrix}$, $T(\mathbf{e_2}) = \begin{bmatrix} -1 \\ 1 \\ 0 \end{bmatrix}$, and $T(\mathbf{e_3}) = \begin{bmatrix} 2 \\ -1 \\ 5 \end{bmatrix}$ are three linear independent vectors in $\mathbb{R}^3$, they form a basis.

17. Is a basis.

19. Is a basis.

21. a. Since $\det(A) = \det\left(\begin{bmatrix} 1 & 0 \\ -2 & -3 \end{bmatrix}\right) = -3 \neq 0$, then the matrix A is invertible and hence, T is an isomorphism. **b.** $A^{-1} = -\frac{1}{3}\begin{bmatrix} -3 & 0 \\ 2 & 1 \end{bmatrix}$ **c.** Let $\mathbf{w} = \begin{bmatrix} x \\ y \end{bmatrix}$. To show that $T^{-1}(\mathbf{w}) = A^{-1}\mathbf{w}$, we will show that $A^{-1}(T(\mathbf{w})) = \mathbf{w}$. That is,

$$A^{-1}T\left(\begin{bmatrix} x \\ y \end{bmatrix}\right) = \begin{bmatrix} 1 & 0 \\ -2/3 & -1/3 \end{bmatrix}\begin{bmatrix} x \\ -2x - 3y \end{bmatrix} = \begin{bmatrix} x \\ y \end{bmatrix}.$$

23. a. Since $\det(A) = \det\left(\begin{bmatrix} -2 & 0 & 1 \\ 1 & -1 & -1 \\ 0 & 1 & 0 \end{bmatrix}\right) = -1 \neq 0$, then T is an isomorphism.

b. $A^{-1} = \begin{bmatrix} -1 & -1 & -1 \\ 0 & 0 & 1 \\ -1 & -2 & -2 \end{bmatrix}$

c. $A^{-1}T\left(\begin{bmatrix} x \\ y \\ z \end{bmatrix}\right) = \begin{bmatrix} -1 & -1 & -1 \\ 0 & 0 & 1 \\ -1 & -2 & -2 \end{bmatrix}\begin{bmatrix} -2x + z \\ x - y - z \\ y \end{bmatrix} = \begin{bmatrix} x \\ y \\ z \end{bmatrix}.$

25. The matrix mapping T is an isomorphism.

27. The matrix mapping T is an isomorphism.

29. Since $T(cA + B) = (cA + B)^t = cA^t + B^t = cT(A) + T(B)$, T is linear. Since $T(A) = \mathbf{0}$ if and only if $A = \mathbf{0}$, then the null space T is $\{\mathbf{0}\}$, so T is one-to-one. To show that T is onto let B be a matrix in $M_{n\times n}$. If $A = B^t$, then $T(A) = T(B^t) = (B^t)^t = B$, so T is onto. Hence, T is an isomorphism.

31. Since $T(kB + C) = A(kB + C)A^{-1} = kABA^{-1} + ACA^{-1} = kT(B) + T(C)$, then T is linear. Since $T(B) = ABA^{-1} = \mathbf{0}$ if and only if $B = \mathbf{0}$ and hence, T is one-to-one. If C is a matrix in $M_{n\times n}$ and $B = A^{-1}CA$, then $T(B) = T(A^{-1}CA) = A(A^{-1}CA)A^{-1} = C$, so T is onto. Hence T is an isomorphism.

33. Define an isomorphism $T : \mathbb{R}^4 \to P_3$, by $T\left(\begin{bmatrix} a \\ b \\ c \\ d \end{bmatrix}\right) = ax^3 + bx^2 + cx + d.$

35. Since the vector space is given by $V = \left\{ \begin{bmatrix} x \\ y \\ x+2y \end{bmatrix} \middle| x,y \in \mathbb{R} \right\}$ define an isomorphism $T : V \to \mathbb{R}^2$ by

$$T\left(\begin{bmatrix} x \\ y \\ x+2y \end{bmatrix}\right) = \begin{bmatrix} x \\ y \end{bmatrix}.$$

37. Let $\mathbf{v}$ be a nonzero vector in $\mathbb{R}^3$. Then a line L through the origin in the direction of the vector $\mathbf{v}$ is given by all scalar multiples of the vector $\mathbf{v}$. That is, $L = \{t\mathbf{v}|\ t \in \mathbb{R}\}$. Now, let $T : \mathbb{R}^3 \longrightarrow \mathbb{R}^3$ be an isomorphism. Since T is linear, then $T(t\mathbf{v}) = tT(\mathbf{v})$. Also, by Theorem 8, $T(\mathbf{v})$ is nonzero. Hence, the set $L' = \{tT(\mathbf{v})|\ t \in \mathbb{R}\}$ is also a line in $\mathbb{R}^3$ through the origin. A plane P is given by the span of two linearly independent vectors $\mathbf{u}$ and $\mathbf{v}$. That is, $P = \{s\mathbf{u} + t\mathbf{v}|\ s,t \in \mathbb{R}\}$. Then $T(s\mathbf{u}+t\mathbf{v}) = sT(\mathbf{u})+tT(\mathbf{v})$, and since T is an isomorphism $T(\mathbf{u})$ and $T(\mathbf{v})$ are linearly independent and hence, $P' = T(P) = \{sT(\mathbf{u})+tT(\mathbf{v})|\ s,t \in \mathbb{R}\}$ is a plane.

Exercise Set 4.4

If A is an $m \times n$ matrix, a mapping $T : \mathbb{R}^n \longrightarrow \mathbb{R}^m$ defined by the matrix product $T(\mathbf{v}) = A\mathbf{v}$ is a linear transformation. In Section 4.4, it is shown how every linear transformation $T : V \longrightarrow W$ can be described by a matrix product. The matrix representation is given relative to bases for the vector spaces V and W and is defined using coordinates relative to these bases. If $B = \{\mathbf{v_1}, \ldots, \mathbf{v_n}\}$ is a basis for V and B' a basis for W, two results are essential in solving the exercises:

- The matrix representation of T relative to B and B' is defined by

$$[T]_B^{B'} = [\ [T(\mathbf{v_1})]_{B'}\ [T(\mathbf{v_2})]_{B'}\ \cdots\ [T(\mathbf{v_n})]_{B'}\ ].$$

- Coordinates of $T(\mathbf{v})$ can be found using the formula

$$[T(\mathbf{v}))]_{B'} = [T]_B^{B'}[\mathbf{v}]_B.$$

To outline the steps required in finding and using a matrix representation of a linear transformation define $T : \mathbb{R}^3 \longrightarrow \mathbb{R}^3$ by $T\left(\begin{bmatrix} x \\ y \\ z \end{bmatrix}\right) = \begin{bmatrix} -x \\ -y \\ z \end{bmatrix}$ and let

$$B = \left\{ \begin{bmatrix} 1 \\ 0 \\ 0 \end{bmatrix}, \begin{bmatrix} 0 \\ 1 \\ 0 \end{bmatrix}, \begin{bmatrix} 0 \\ 0 \\ 1 \end{bmatrix} \right\} \quad \text{and} \quad B' = \left\{ \begin{bmatrix} 1 \\ 1 \\ 1 \end{bmatrix}, \begin{bmatrix} 1 \\ 0 \\ 1 \end{bmatrix}, \begin{bmatrix} 2 \\ 1 \\ 0 \end{bmatrix} \right\}$$

two bases for $\mathbb{R}^3$.

- Apply T to each basis vector in B.

$$T\left(\begin{bmatrix} 1 \\ 0 \\ 0 \end{bmatrix}\right) = \begin{bmatrix} -1 \\ 0 \\ 0 \end{bmatrix}, T\left(\begin{bmatrix} 0 \\ 1 \\ 0 \end{bmatrix}\right) = \begin{bmatrix} 0 \\ -1 \\ 0 \end{bmatrix}, T\left(\begin{bmatrix} 0 \\ 0 \\ 1 \end{bmatrix}\right) = \begin{bmatrix} 0 \\ 0 \\ 1 \end{bmatrix}$$

- Find the coordinates of each of the vectors found in the first step relative to B'. Since

$$\begin{bmatrix} 1 & 1 & 2 & | & -1 & 0 & 0 \\ 1 & 0 & 1 & | & 0 & -1 & 0 \\ 1 & 1 & 0 & | & 0 & 0 & 1 \end{bmatrix} \longrightarrow \begin{bmatrix} 1 & 0 & 0 & | & 1/2 & -1 & 1/2 \\ 1 & 0 & 1 & | & -1/2 & 1 & 1/2 \\ 1 & 1 & 0 & | & -1/2 & 0 & -1/2 \end{bmatrix},$$

then

$$\left[\begin{bmatrix} -1 \\ 0 \\ 0 \end{bmatrix}\right]_{B'} = \begin{bmatrix} 1/2 \\ -1/2 \\ -1/2 \end{bmatrix}, \left[\begin{bmatrix} 0 \\ -1 \\ 0 \end{bmatrix}\right]_{B'} = \begin{bmatrix} -1 \\ 1 \\ 0 \end{bmatrix}, \left[\begin{bmatrix} 0 \\ 0 \\ 1 \end{bmatrix}\right]_{B'} = \begin{bmatrix} 1/2 \\ 1/2 \\ -1/2 \end{bmatrix}.$$

- The column vectors of the matrix representation relative to B and B' are the coordinate vectors found in the previous step.

$$[T]_B^{B'} = \begin{bmatrix} 1/2 & -1 & 1/2 \\ -1/2 & 1 & 1/2 \\ -1/2 & 0 & -1/2 \end{bmatrix}$$

- The coordinates of any vector $T(\mathbf{v})$ can be found using the matrix product

$$[T(\mathbf{v}))]_{B'} = [T]_B^{B'}[\mathbf{v}]_B.$$

- As an example, let $\mathbf{v} = \begin{bmatrix} 1 \\ -2 \\ -4 \end{bmatrix}$, then after applying the operator T the coordinates relative to B' is given by

$$\left[T\left(\begin{bmatrix} 1 \\ -2 \\ -4 \end{bmatrix} \right) \right]_{B'} = \begin{bmatrix} 1/2 & -1 & 1/2 \\ -1/2 & 1 & 1/2 \\ -1/2 & 0 & -1/2 \end{bmatrix} \left[\begin{bmatrix} 1 \\ -2 \\ -4 \end{bmatrix} \right]_B.$$

Since B is the standard basis the coordinates of a vector are just the components, so

$$\left[T\left(\begin{bmatrix} 1 \\ -2 \\ -4 \end{bmatrix} \right) \right]_{B'} = \begin{bmatrix} 1/2 & -1 & 1/2 \\ -1/2 & 1 & 1/2 \\ -1/2 & 0 & -1/2 \end{bmatrix} \begin{bmatrix} 1 \\ -2 \\ -4 \end{bmatrix} = \begin{bmatrix} 1/2 \\ -9/2 \\ 3/2 \end{bmatrix}.$$

This vector is not $T(\mathbf{v})$, but the coordinates relative to the basis B'. Then

$$T\left(\begin{bmatrix} 1 \\ -2 \\ -4 \end{bmatrix} \right) = \frac{1}{2}\begin{bmatrix} 1 \\ 1 \\ 1 \end{bmatrix} - \frac{9}{2}\begin{bmatrix} 1 \\ 0 \\ 1 \end{bmatrix} + \frac{3}{2}\begin{bmatrix} 2 \\ 1 \\ 0 \end{bmatrix} = \begin{bmatrix} -1 \\ 2 \\ -7/2 \end{bmatrix}.$$

Other useful formulas that involve combinations of linear transformations and the matrix representation are:

- $[S+T]_B^{B'} = [S]_B^{B'} + [T]_B^{B'}$ • $[kT]_B^{B'} = k[T]_B^{B'}$ • $[S \circ T]_B^{B'} = [S]_B^{B'}[T]_B^{B'}$ • $[T^n]_B = ([T]_B)^n$ • $[T^{-1}]_B = ([T]_B)^{-1}$

■ Solutions to Odd Exercises

1. a. Let $B = \{\mathbf{e_1}, \mathbf{e_2}\}$ be the standard basis. To find the matrix representation for A relative to B, the column vectors are the coordinates of $T(\mathbf{e_1})$ and $T(\mathbf{e_2})$ relative to B. Recall the coordinates of a vector relative to the standard basis are just the components of the vector. Hence, $[T]_B = [\ [T(\mathbf{e_1})]_B\ [T(\mathbf{e_2})]_B\ ] = \begin{bmatrix} 5 & -1 \\ -1 & 1 \end{bmatrix}$. **b.** The direct computation is $T\begin{bmatrix} 2 \\ 1 \end{bmatrix} = \begin{bmatrix} 9 \\ -1 \end{bmatrix}$ and using part (a), the result is $T\begin{bmatrix} 2 \\ 1 \end{bmatrix} = \begin{bmatrix} 5 & -1 \\ -1 & 1 \end{bmatrix}\begin{bmatrix} 2 \\ 1 \end{bmatrix} = \begin{bmatrix} 9 \\ -1 \end{bmatrix}$.

3. a. Let $B = \{\mathbf{e_1}, \mathbf{e_2}, \mathbf{e_3}\}$ be the standard basis. Then $[T]_B = [\ [T(\mathbf{e_1})]_B\ [T(\mathbf{e_2})]_B\ [T(\mathbf{e_2})]_B\ ] = \begin{bmatrix} -1 & 1 & 2 \\ 0 & 3 & 1 \\ 1 & 0 & -1 \end{bmatrix}$. **b.** The direct computation is $T\begin{bmatrix} 1 \\ -2 \\ 3 \end{bmatrix} = \begin{bmatrix} 3 \\ -3 \\ -2 \end{bmatrix}$, and using part (a) the result is

$$T\begin{bmatrix} 1 \\ -2 \\ 3 \end{bmatrix} = \begin{bmatrix} -1 & 1 & 2 \\ 0 & 3 & 1 \\ 1 & 0 & -1 \end{bmatrix}\begin{bmatrix} 1 \\ -2 \\ 3 \end{bmatrix} = \begin{bmatrix} 3 \\ -3 \\ -2 \end{bmatrix}.$$

5. a. The column vectors of the matrix representation relative to B and B' are the coordinates relative to B' of the images of the vectors in B by T. That is, $[T]_B^{B'} = \left[\left[T\left(\begin{bmatrix} 1 \\ -1 \end{bmatrix} \right) \right]_{B'} \; \left[T\left(\begin{bmatrix} 2 \\ 0 \end{bmatrix} \right) \right]_{B'} \right]$. Since B' is the standard basis, the coordinates are the components of the vectors $T\left(\begin{bmatrix} 1 \\ -1 \end{bmatrix} \right)$ and $T\left(\begin{bmatrix} 2 \\ 0 \end{bmatrix} \right)$, so $[T]_B^{B'} = \begin{bmatrix} -3 & -2 \\ 3 & 6 \end{bmatrix}$.

b. The direct computation is $T\begin{bmatrix} -1 \\ -2 \end{bmatrix} = \begin{bmatrix} -3 \\ -3 \end{bmatrix}$ and using part (a)

$$T\begin{bmatrix} -1 \\ -2 \end{bmatrix} = \begin{bmatrix} -3 & -2 \\ 3 & 6 \end{bmatrix} \begin{bmatrix} -1 \\ -2 \end{bmatrix}_B = \begin{bmatrix} -3 & -2 \\ 3 & 6 \end{bmatrix} \begin{bmatrix} 2 \\ -3/2 \end{bmatrix} = \begin{bmatrix} -3 \\ -3 \end{bmatrix}.$$

7. a. The matrix representation is given by

$$[T]_B^{B'} = \left[\left[T\left(\begin{bmatrix} -1 \\ -2 \end{bmatrix} \right) \right]_{B'} \; \left[T\left(\begin{bmatrix} 1 \\ 1 \end{bmatrix} \right) \right]_{B'} \right] = \left[\left[T\left(\begin{bmatrix} -2 \\ -3 \end{bmatrix} \right) \right]_{B'} \; \left[T\left(\begin{bmatrix} 2 \\ 2 \end{bmatrix} \right) \right]_{B'} \right].$$

We can find the coordinates of both vectors by considering

$$\left[\begin{array}{cc|cc} 3 & 0 & -2 & 2 \\ -2 & -2 & -3 & 2 \end{array} \right] \longrightarrow \left[\begin{array}{cc|cc} 1 & 0 & -\frac{2}{3} & \frac{2}{3} \\ 0 & 1 & \frac{13}{6} & -\frac{5}{3} \end{array} \right], \text{ so } [T]_B^{B'} = \begin{bmatrix} -\frac{2}{3} & \frac{2}{3} \\ \frac{13}{6} & -\frac{5}{3} \end{bmatrix}.$$

b. The direct computation is $T\begin{bmatrix} -1 \\ -3 \end{bmatrix} = \begin{bmatrix} -2 \\ -4 \end{bmatrix}$. Using part (a) we can now find the coordinates of the image of a vector using the formula $[T(\mathbf{v})]_{B'} = [T]_B^{B'}[\mathbf{v}]_B$. and then use these coordinates to fine $T(\mathbf{v})$. That is,

$$\left[T\begin{bmatrix} -1 \\ -3 \end{bmatrix} \right]_{B'} = [T]_B^{B'} \begin{bmatrix} -1 \\ -3 \end{bmatrix}_B = [T]_B^{B'} \begin{bmatrix} 2 \\ 1 \end{bmatrix} = \begin{bmatrix} -\frac{2}{3} \\ \frac{8}{3} \end{bmatrix}, \text{ so } T\begin{bmatrix} -1 \\ -3 \end{bmatrix} = -\frac{2}{3}\begin{bmatrix} 3 \\ -2 \end{bmatrix} + \frac{8}{3}\begin{bmatrix} 0 \\ -2 \end{bmatrix} = \begin{bmatrix} -2 \\ -4 \end{bmatrix}.$$

9. a. Since B' is the standard basis for $\mathcal{P}_2$, then $[T]_B^{B'} = \begin{bmatrix} 1 & 1 & 1 \\ 0 & -1 & -2 \\ 0 & 0 & 1 \end{bmatrix}$. **b.** The direct computation is $T(x^2 - 3x + 3) = x^2 - 3x + 3$. To find the coordinates of the image, we have from part (a) that

$$\left[T(x^2 - 3x + 3) \right]_{B'} = [T]_B^{B'}[x^2 - 3x + 3]_B = [T]_B^{B'} \begin{bmatrix} 1 \\ 1 \\ 1 \end{bmatrix} = \begin{bmatrix} 3 \\ -3 \\ 1 \end{bmatrix}, \text{ so } T(x^2 - 3x + 3) = 3 - 3x + x^2.$$

11. First notice that if $A = \begin{bmatrix} a & b \\ c & -a \end{bmatrix}$, then $T(A) = \begin{bmatrix} 0 & -2b \\ 2c & 0 \end{bmatrix}$.

a. $[T]_B = \begin{bmatrix} 0 & 0 & 0 \\ 0 & -2 & 0 \\ 0 & 0 & 2 \end{bmatrix}$ **b.** The direct computation is $T\left(\begin{bmatrix} 2 & 1 \\ 3 & -2 \end{bmatrix} \right) = \begin{bmatrix} 0 & -2 \\ 6 & 0 \end{bmatrix}$. Using part (a)

$$\left[T\left(\begin{bmatrix} 2 & 1 \\ 3 & -2 \end{bmatrix} \right) \right]_B = [T]_B \begin{bmatrix} 2 \\ 1 \\ 3 \end{bmatrix} = \begin{bmatrix} 0 \\ -2 \\ 6 \end{bmatrix}$$

so

$$T\left(\begin{bmatrix} 2 & 1 \\ 3 & -2 \end{bmatrix} \right) = 0\begin{bmatrix} 1 & 0 \\ 0 & -1 \end{bmatrix} - 2\begin{bmatrix} 0 & 1 \\ 0 & 0 \end{bmatrix} + 6\begin{bmatrix} 0 & 0 \\ 1 & 0 \end{bmatrix} = \begin{bmatrix} 0 & -2 \\ 6 & 0 \end{bmatrix}.$$

13. a. $[T]_B = \begin{bmatrix} 1 & 2 \\ 1 & -1 \end{bmatrix}$ **b.** $[T]_{B'} = \frac{1}{9}\begin{bmatrix} 1 & 22 \\ 11 & -1 \end{bmatrix}$ **c.** $[T]_B^{B'} = \frac{1}{9}\begin{bmatrix} 5 & -2 \\ 1 & 5 \end{bmatrix}$

d. $[T]_{B'}^B = \frac{1}{3}\begin{bmatrix} 5 & 2 \\ -1 & 5 \end{bmatrix}$ **e.** $[T]_C^{B'} = \frac{1}{9}\begin{bmatrix} -2 & 5 \\ 5 & 1 \end{bmatrix}$ **f.** $[T]_{C'}^{B'} = \frac{1}{9}\begin{bmatrix} 22 & 1 \\ -1 & 11 \end{bmatrix}$

15. a. $[T]_B^{B'} = \begin{bmatrix} 0 & 0 \\ 1 & 0 \\ 0 & 1/2 \end{bmatrix}$ **b.** $[T]_C^{B'} = \begin{bmatrix} 0 & 0 \\ 0 & 1 \\ 1/2 & 0 \end{bmatrix}$ **c.** $[T]_C^{C'} = \begin{bmatrix} 0 & 1 \\ 0 & 0 \\ 1/2 & 0 \end{bmatrix}$ **d.** $[S]_{B'}^B = \begin{bmatrix} 0 & 1 & 0 \\ 0 & 0 & 2 \end{bmatrix}$

e. $[S]_{B'}^B[T]_B^{B'} = \begin{bmatrix} 1 & 0 \\ 0 & 1 \end{bmatrix}, [T]_B^{B'}[S]_{B'}^B = \begin{bmatrix} 0 & 0 & 0 \\ 0 & 1 & 0 \\ 0 & 0 & 1 \end{bmatrix}$ **f.** The function $S \circ T$ is the identity map, that is,

$(S \circ T)(ax + b) = ax + b$ so S reverses the action of T.

17. $[T]_B = \begin{bmatrix} 1 & 0 \\ 0 & -1 \end{bmatrix}$. The transformation T reflects a vector across the x-axis.

19. $[T]_B = cI$ **21.** $[T]_B^{B'} = [1\ 0\ 0\ 1]$

23. a. $[2T + S]_B = 2[T]_B + [S]_B = \begin{bmatrix} 5 & 2 \\ -1 & 7 \end{bmatrix}$ **25. a.** $[S \circ T]_B = [S]_B[T]_B = \begin{bmatrix} 2 & 1 \\ 1 & 4 \end{bmatrix}$ **b.**

b. $\begin{bmatrix} -4 \\ 23 \end{bmatrix}$ $\begin{bmatrix} -1 \\ 10 \end{bmatrix}$

27. a. $[-3T + 2S]_B = \begin{bmatrix} 3 & 3 & 1 \\ 2 & -6 & -6 \\ 3 & -3 & -1 \end{bmatrix}$ **29. a.** $[S \circ T]_B = \begin{bmatrix} 4 & -4 & -4 \\ 1 & -1 & -1 \\ -1 & 1 & 1 \end{bmatrix}$ **b.**

b. $\begin{bmatrix} 3 \\ -26 \\ -9 \end{bmatrix}$ $\begin{bmatrix} -20 \\ -5 \\ 5 \end{bmatrix}$

31. $[T]_B = \begin{bmatrix} 0 & 0 & 0 & 6 & 0 \\ 0 & 0 & 0 & 0 & 24 \\ 0 & 0 & 0 & 0 & 0 \\ 0 & 0 & 0 & 0 & 0 \\ 0 & 0 & 0 & 0 & 0 \end{bmatrix};$ **33.** $[S]_B^{B'} = \begin{bmatrix} 0 & 0 & 0 \\ 1 & 0 & 0 \\ 0 & 1 & 0 \\ 0 & 0 & 1 \end{bmatrix};$

$[T(p(x))]_B = \begin{bmatrix} -12 \\ -48 \\ 0 \\ 0 \\ 0 \end{bmatrix};$ $[D]_B^{B'} = \begin{bmatrix} 0 & 1 & 0 & 0 \\ 0 & 0 & 2 & 0 \\ 0 & 0 & 0 & 3 \end{bmatrix};$

$T(p(x)) = p'''(x) = -12 - 48x$ $[D]_B^{B'}[S]_B^{B'} = \begin{bmatrix} 1 & 0 & 0 \\ 0 & 2 & 0 \\ 0 & 0 & 3 \end{bmatrix} = [T]_B$

35. If $A = \begin{bmatrix} a & b \\ c & d \end{bmatrix}$, then the matrix representation for T is $[T]_S = \begin{bmatrix} 0 & -c & b & 0 \\ -b & a-d & 0 & b \\ c & 0 & d-a & -c \\ 0 & c & -b & 0 \end{bmatrix}.$

37.

$[T]_B = [\ [T(\mathbf{v_1})]_B\ [T(\mathbf{v_2})]_B \ldots [T(\mathbf{v_n})\ ]_B = [\ [\mathbf{v_1}]_B\ [\mathbf{v_1} + \mathbf{v_2}]_B \ldots [\mathbf{v_{n-1}} + \mathbf{v_n}\ ]_B\ ] = \begin{bmatrix} 1 & 1 & 0 & 0 & \ldots & \ldots & 0 & 0 \\ 0 & 1 & 1 & 0 & \ldots & \ldots & 0 & 0 \\ 0 & 0 & 1 & 1 & \ldots & \ldots & 0 & 0 \\ \vdots & \vdots & \vdots & \vdots & \vdots & \vdots & \vdots & \vdots \\ 0 & 0 & 0 & 0 & 0 & \ldots & 1 & 0 \\ 0 & 0 & 0 & 0 & 0 & \ldots & 1 & 1 \\ 0 & 0 & 0 & 0 & 0 & \ldots & 0 & 1 \end{bmatrix}$

Exercise Set 4.5

If $T : V \longrightarrow V$ is a linear operator the matrix representation of T relative to a basis B, denoted $[T]_B$ depends on the basis. However, the action of the operator does not change, so does not depend on the matrix representation. Suppose $B_1 = \{\mathbf{v_1}, \ldots, \mathbf{v_n}\}$ and $B_2\{\mathbf{v'_1}, \ldots, \mathbf{v'_n}\}$ are two bases for V and the coordinates of $T(\mathbf{v})$ are $[T(\mathbf{v})]_{B_1} = \begin{bmatrix} c_1 \\ c_2 \\ \vdots \\ c_n \end{bmatrix}$ and $[T(\mathbf{v})]_{B_2} = \begin{bmatrix} d_1 \\ d_2 \\ \vdots \\ d_n \end{bmatrix}$. Then

$$T(\mathbf{v}) = c_1\mathbf{v_1} + c_2\mathbf{v_2} + \cdots + c_n\mathbf{v_n} = d_1\mathbf{v'_1} + d_2\mathbf{v'_2} + \cdots + d_n\mathbf{v'_n}.$$

The matrix representations are related by the formula

$$[T]_{B_2} = [I]_{B_1}^{B_2}[T]_{B_1}[I]_{B_2}^{B_1}.$$

Recall that transition matrices are invertible with $[I]_{B_1}^{B_2} = ([I]_{B_2}^{B_1})^{-1}$. Two $n \times n$ matrices A and B are called similar provided there exists an invertible matrix P such that $B = P^{-1}AP$.

■ Solutions to Odd Exercises

1. The coordinates of the image of the vector $\mathbf{v} = \begin{bmatrix} 4 \\ -1 \end{bmatrix}$ relative to the two bases are

$$[T(\mathbf{v})]_{B_1} = [T]_{B_1}[\mathbf{v}]_{B_1} = \begin{bmatrix} 1 & 2 \\ -1 & 3 \end{bmatrix}\begin{bmatrix} 4 \\ -1 \end{bmatrix} = \begin{bmatrix} 2 \\ -7 \end{bmatrix} \text{ and } [T(\mathbf{v})]_{B_2} = [T]_{B_2}[\mathbf{v}]_{B_2} = \begin{bmatrix} 2 & 1 \\ -1 & 2 \end{bmatrix}\begin{bmatrix} -1 \\ -5 \end{bmatrix} = \begin{bmatrix} -7 \\ -9 \end{bmatrix}$$

Then using the coordinates relative to the respective bases the vector $T(\mathbf{v})$ is

$$-7\begin{bmatrix} 1 \\ 1 \end{bmatrix} + (-9)\begin{bmatrix} -1 \\ 0 \end{bmatrix} = \begin{bmatrix} 2 \\ -7 \end{bmatrix} = 2\begin{bmatrix} 1 \\ 0 \end{bmatrix} - 7\begin{bmatrix} 0 \\ 1 \end{bmatrix},$$

so the action of the operator is the same regardless of the particular basis used.

3. a. Since B_1 is the standard basis, $T(\mathbf{e_1}) = \begin{bmatrix} 1 \\ 1 \end{bmatrix}$, and $T(\mathbf{e_2}) = \begin{bmatrix} 1 \\ 1 \end{bmatrix}$, then $[T]_{B_1} = \begin{bmatrix} 1 & 1 \\ 1 & 1 \end{bmatrix}$. Relative to the basis B_2, $T\left(\begin{bmatrix} 1 \\ 1 \end{bmatrix}\right) = \begin{bmatrix} 2 \\ 2 \end{bmatrix} = 2\begin{bmatrix} 1 \\ 1 \end{bmatrix} + 0\begin{bmatrix} -1 \\ 1 \end{bmatrix}$, and $T\left(\begin{bmatrix} -1 \\ 1 \end{bmatrix}\right) = \begin{bmatrix} 0 \\ 0 \end{bmatrix}$, so $[T]_{B_2} = \begin{bmatrix} 2 & 0 \\ 0 & 0 \end{bmatrix}$.

b. The coordinates of the image of the vector $\mathbf{v} = \begin{bmatrix} 3 \\ -2 \end{bmatrix}$ relative to the two bases are

$$[T]_{B_1}[\mathbf{v}]_{B_1} = \begin{bmatrix} 1 & 1 \\ 1 & 1 \end{bmatrix}\begin{bmatrix} 3 \\ -2 \end{bmatrix} = \begin{bmatrix} 1 \\ 1 \end{bmatrix} \text{ and } [T]_{B_2}[\mathbf{v}]_{B_2} = \begin{bmatrix} 2 & 0 \\ 0 & 0 \end{bmatrix}\begin{bmatrix} 1/2 \\ -5/2 \end{bmatrix} = \begin{bmatrix} 1 \\ 0 \end{bmatrix}.$$

Then using the coordinates relative to the respective bases the vector $T(\mathbf{v})$ is

$$1\begin{bmatrix} 1 \\ 1 \end{bmatrix} + (0)\begin{bmatrix} -1 \\ 1 \end{bmatrix} = \begin{bmatrix} 1 \\ 1 \end{bmatrix} = \begin{bmatrix} 1 \\ 0 \end{bmatrix} + \begin{bmatrix} 0 \\ 1 \end{bmatrix},$$

so the action of the operator is the same regardless of the particular basis used.

5. a. $[T]_{B_1} = \begin{bmatrix} 1 & 0 & 0 \\ 0 & 0 & 0 \\ 0 & 0 & 1 \end{bmatrix}$, $[T]_{B_2} = \begin{bmatrix} 1 & -1 & 0 \\ 0 & 0 & 0 \\ 0 & 1 & 1 \end{bmatrix}$

b. $[T]_{B_1}[\mathbf{v}]_{B_1} = \begin{bmatrix} 1 & 0 & 0 \\ 0 & 0 & 0 \\ 0 & 0 & 1 \end{bmatrix} \begin{bmatrix} 1 \\ 2 \\ -1 \end{bmatrix} = \begin{bmatrix} 1 \\ 0 \\ -1 \end{bmatrix}$, $[T]_{B_2}[\mathbf{v}]_{B_2} = \begin{bmatrix} 1 & -1 & 0 \\ 0 & 0 & 0 \\ 0 & 1 & 1 \end{bmatrix} \begin{bmatrix} 3 \\ 2 \\ -4 \end{bmatrix} = \begin{bmatrix} 1 \\ 0 \\ -2 \end{bmatrix}$. Since

B_1 is the standard basis and $1\begin{bmatrix} 1 \\ 0 \\ 1 \end{bmatrix} + (0)\begin{bmatrix} -1 \\ 1 \\ 0 \end{bmatrix} + (-2)\begin{bmatrix} 0 \\ 0 \\ 1 \end{bmatrix} = \begin{bmatrix} 1 \\ 0 \\ -1 \end{bmatrix}$, the action of the operator is

the same regardless of the particular basis used.

7. Since B_1 is the standard basis, then the troansition matrix relative to B_2 and B_1 is

$P = [I]_{B_2}^{B_1} = \left[\left[\begin{bmatrix} 3 \\ -1 \end{bmatrix} \right]_{B_1} \left[\begin{bmatrix} -1 \\ 1 \end{bmatrix} \right]_{B_1} \right] = \begin{bmatrix} 3 & -1 \\ -1 & 1 \end{bmatrix}$. By Theorem 15,

$$[T]_{B_2} = P^{-1}[T]_{B_1}P = \frac{1}{2}\begin{bmatrix} 1 & 1 \\ 1 & 3 \end{bmatrix}\begin{bmatrix} 1 & 1 \\ 3 & 2 \end{bmatrix}\begin{bmatrix} 3 & -1 \\ -1 & 1 \end{bmatrix} = \begin{bmatrix} 9/2 & -1/2 \\ 23/2 & -3/2 \end{bmatrix}.$$

9. The troansition matrix is $P = [I]_{B_2}^{B_1} = \begin{bmatrix} 1/3 & 1 \\ 1/3 & -1 \end{bmatrix}$. By Theorem 15

$$[T]_{B_2} = P^{-1}[T]_{B_1}P = \begin{bmatrix} 3/2 & 3/2 \\ 1/2 & -1/2 \end{bmatrix}\begin{bmatrix} 1 & 0 \\ 0 & -1 \end{bmatrix}\begin{bmatrix} 1/3 & 1 \\ 1/3 & -1 \end{bmatrix} = \begin{bmatrix} 0 & 3 \\ 1/3 & 0 \end{bmatrix}.$$

11. Since $P = [I]_{B_2}^{B_1} = \begin{bmatrix} 2 & 1 \\ 3 & 2 \end{bmatrix}$ and $[T]_{B_1} = \begin{bmatrix} 2 & 0 \\ 0 & 0 \end{bmatrix}$, by Theorem 15,

$$[T]_{B_2} = P^{-1}[T]_{B_1}P = \begin{bmatrix} 2 & -1 \\ -3 & 2 \end{bmatrix}\begin{bmatrix} 2 & 0 \\ 0 & 3 \end{bmatrix}\begin{bmatrix} 2 & 1 \\ 3 & 2 \end{bmatrix} = \begin{bmatrix} -1 & -2 \\ 6 & 6 \end{bmatrix}.$$

13. Since $P = [I]_{B_2}^{B_1} = \begin{bmatrix} -1 & -1 \\ 2 & 1 \end{bmatrix}$, and $[T]_{B_1} = \begin{bmatrix} 1 & -1 \\ -2 & 1 \end{bmatrix}$, by Theorem 15

$$[T]_{B_2} = P^{-1}[T]_{B_1}P = \begin{bmatrix} 1 & 1 \\ -2 & -1 \end{bmatrix}\begin{bmatrix} 1 & -1 \\ -2 & 1 \end{bmatrix}\begin{bmatrix} -1 & -1 \\ 2 & 1 \end{bmatrix} = \begin{bmatrix} 1 & 1 \\ 2 & 1 \end{bmatrix}.$$

15. Since $T(1) = 0, T(x) = 1$, and $T(x^2) = 2x$, then $[T]_{B_1} = \begin{bmatrix} 0 & 1 & 0 \\ 0 & 0 & 2 \\ 0 & 0 & 0 \end{bmatrix}$ and

$[T]_{B_2} = \begin{bmatrix} 0 & 2 & 0 \\ 0 & 0 & 1 \\ 0 & 0 & 0 \end{bmatrix}$. Now if $P = [I]_{B_2}^{B_1} = \begin{bmatrix} 1 & 0 & -2 \\ 0 & 2 & 0 \\ 0 & 0 & 1 \end{bmatrix}$, then by Theorem 15, $[T]_{B_2} = P^{-1}[T]_{B_1}P$.

17. Since A and B are similar, there is an invertible matrix P such that $B = P^{-1}AP$. Also since B and C are similar, there is an invertible matrix Q such that $C = Q^{-1}BQ$. Therefore, $C = Q^{-1}P^{-1}APQ = (PQ)^{-1}A(PQ)$, so that A and C are also similar.

19. For any square matrices A and B, the trace function satisfies the property $\mathbf{tr}(AB) = \mathbf{tr}(BA)$. Now, since A and B are similar matrices there exists an invertible matrix P such that $B = P^{-1}AP$. Hence

$$\mathbf{tr}(B) = \mathbf{tr}(P^{-1}AP) = \mathbf{tr}(APP^{-1}) = \mathbf{tr}(A).$$

21. Since A and B are similar matrices there exists an invertible matrix P such that $B = P^{-1}AP$. Hence

$$B^n = (P^{-1}AP)^n = P^{-1}A^nP.$$

Thus, A^n and B^n are similar.

Exercise Set 4.6

1. a. Since the triangle is reflected across the x-axis, the matrix representation relative to the standard basis for T is $\begin{bmatrix} 1 & 0 \\ 0 & -1 \end{bmatrix}$. **b.** Since the triangle is reflected across the y-axis, the matrix representation relative to the standard basis is $\begin{bmatrix} -1 & 0 \\ 0 & 1 \end{bmatrix}$. **c.** Since the triangle is vertically stretched by a factor of 3, the matrix representation relative to the standard basis is $\begin{bmatrix} 1 & 0 \\ 0 & 3 \end{bmatrix}$.

3. a. The matrix representation relative to the standard basis S is the product of the matrix representations for the three separate operators. That is,

$$[T]_S = \begin{bmatrix} 1 & 0 \\ 0 & -1 \end{bmatrix} \begin{bmatrix} 1 & 0 \\ 0 & \frac{1}{2} \end{bmatrix} \begin{bmatrix} 3 & 0 \\ 0 & 1 \end{bmatrix} = \begin{bmatrix} 3 & 0 \\ 0 & -1/2 \end{bmatrix}.$$

b.

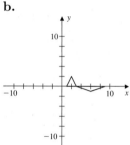

c. The matrix that will reverse the action of the operator T is the inverse of $[T]_S$. That is,

$$[T]_S^{-1} = \begin{bmatrix} 1/3 & 0 \\ 0 & -2 \end{bmatrix}.$$

5. a.

$$[T]_S = \begin{bmatrix} -\sqrt{2}/2 & \sqrt{2}/2 \\ -\sqrt{2}/2 & -\sqrt{2}/2 \end{bmatrix}$$

b.

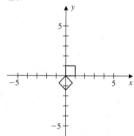

c.

$$[T]_S^{-1} = \begin{bmatrix} -\sqrt{2}/2 & -\sqrt{2}/2 \\ \sqrt{2}/2 & -\sqrt{2}/2 \end{bmatrix}$$

7. a. $[T]_S = \begin{bmatrix} \sqrt{3}/2 & -1/2 & 0 \\ 1/2 & \sqrt{3}/2 & 0 \\ 0 & 0 & 1 \end{bmatrix} \begin{bmatrix} 1 & 0 & 1 \\ 0 & 1 & 1 \\ 0 & 0 & 1 \end{bmatrix} = \begin{bmatrix} \sqrt{3}/2 & -1/2 & \sqrt{3}/2 - 1/2 \\ 1/2 & \sqrt{3}/2 & \sqrt{3}/2 + 1/2 \\ 0 & 0 & 1 \end{bmatrix}$

c. $\begin{bmatrix} \sqrt{3}/2 & 1/2 & -1 \\ -1/2 & \sqrt{3}/2 & -1 \\ 0 & 0 & 1 \end{bmatrix}$

b.

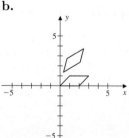

9. a. $\left[\begin{bmatrix} 0 \\ 0 \end{bmatrix} \right]_B = \begin{bmatrix} 0 \\ 0 \end{bmatrix}$, $\left[\begin{bmatrix} 2 \\ 2 \end{bmatrix} \right]_B = \begin{bmatrix} 2 \\ 0 \end{bmatrix}$, $\left[\begin{bmatrix} 0 \\ 2 \end{bmatrix} \right]_B = \begin{bmatrix} 1 \\ 1 \end{bmatrix}$ **b.** Since the operator T is given by $T\left(\begin{bmatrix} x \\ y \end{bmatrix} \right) = \begin{bmatrix} y \\ x \end{bmatrix}$, then $[T]_B^S = \left[\left[T\left(\begin{bmatrix} 1 \\ 1 \end{bmatrix} \right) \right]_S \quad \left[T\left(\begin{bmatrix} -1 \\ 1 \end{bmatrix} \right) \right]_S \right] = \begin{bmatrix} 1 & 1 \\ 1 & -1 \end{bmatrix}.$

c. $\begin{bmatrix} 1 & 1 \\ 1 & -1 \end{bmatrix}\begin{bmatrix} 0 \\ 0 \end{bmatrix} = \begin{bmatrix} 0 \\ 0 \end{bmatrix}$, $\begin{bmatrix} 1 & 1 \\ 1 & -1 \end{bmatrix}\begin{bmatrix} 2 \\ 0 \end{bmatrix} = \begin{bmatrix} 2 \\ 2 \end{bmatrix}$, $\begin{bmatrix} 1 & 1 \\ 1 & -1 \end{bmatrix}\begin{bmatrix} 1 \\ 1 \end{bmatrix} = \begin{bmatrix} 2 \\ 0 \end{bmatrix}$.

The original triangle is reflected across the line $y = x$, as shown in the figure.

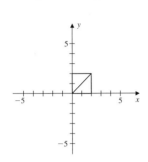

d. $\begin{bmatrix} 0 & 1 \\ 1 & 0 \end{bmatrix}\begin{bmatrix} 0 \\ 0 \end{bmatrix} = \begin{bmatrix} 0 \\ 0 \end{bmatrix}$,

$\begin{bmatrix} 0 & 1 \\ 1 & 0 \end{bmatrix}\begin{bmatrix} 2 \\ 2 \end{bmatrix} = \begin{bmatrix} 2 \\ 2 \end{bmatrix}$,

$\begin{bmatrix} 0 & 1 \\ 1 & 0 \end{bmatrix}\begin{bmatrix} 0 \\ 2 \end{bmatrix} = \begin{bmatrix} 2 \\ 0 \end{bmatrix}$

Review Exercises Chapter 4

1. a. The vectors are not scalar multiples, so S is a basis

b. Since S is a basis, for any vector $\begin{bmatrix} x \\ y \end{bmatrix}$ there are scalars c_1 and c_2 such that

$\begin{bmatrix} x \\ y \end{bmatrix} = c_1 \begin{bmatrix} 1 \\ 1 \end{bmatrix} + c_2 \begin{bmatrix} 3 \\ -1 \end{bmatrix}$. The resulting linear system $\begin{cases} c_1 + 3c_2 &= x \\ c_1 - c_2 &= y \end{cases}$ has the unique solution $c_1 = \frac{1}{4}x + \frac{3}{4}y$ and $c_2 = \frac{1}{4}x - \frac{1}{4}y$. Then

$$T\left(\begin{bmatrix} x \\ y \end{bmatrix}\right) = c_1 T\left(\begin{bmatrix} 1 \\ 1 \end{bmatrix}\right) + c_2 T\left(\begin{bmatrix} 3 \\ -1 \end{bmatrix}\right) = \begin{bmatrix} x \\ x+y \\ x-y \\ 2y \end{bmatrix}.$$

c. $N(T) = \{\mathbf{0}\}$ **d.** Since $N(T) = \{\mathbf{0}\}$, T is one-to-one.

e. Since the range consists of all vectors of the form $\begin{bmatrix} x \\ x+y \\ x-y \\ 2y \end{bmatrix} = x\begin{bmatrix} 1 \\ 1 \\ 1 \\ 2 \end{bmatrix} + y\begin{bmatrix} 0 \\ 1 \\ -1 \\ 2 \end{bmatrix}$ and the vectors $\begin{bmatrix} 1 \\ 1 \\ 1 \\ 2 \end{bmatrix}$

and $\begin{bmatrix} 0 \\ 1 \\ -1 \\ 2 \end{bmatrix}$ are linearly independent, then a basis for the range is $\left\{ \begin{bmatrix} 1 \\ 1 \\ 1 \\ 2 \end{bmatrix}, \begin{bmatrix} 0 \\ 1 \\ -1 \\ 2 \end{bmatrix} \right\}$.

f. Since $\dim(R(T)) = 2$ and $\dim(\mathbb{R}^4) = 4$, then T is not onto. Also $\begin{bmatrix} a \\ b \\ c \\ d \end{bmatrix}$ is in $R(T)$ if and only if $c+b-2a = 0$.

g. $\left\{ \begin{bmatrix} 1 \\ 0 \\ 1 \\ 1 \end{bmatrix}, \begin{bmatrix} -1 \\ 1 \\ 0 \\ 1 \end{bmatrix}, \begin{bmatrix} 1 \\ 0 \\ 0 \\ 0 \end{bmatrix}, \begin{bmatrix} 0 \\ 1 \\ 0 \\ 0 \end{bmatrix} \right\}$ **h.** $[T]_B^C = \begin{bmatrix} -1 & 2 \\ 5 & -4 \\ 7 & -5 \\ -2 & 4 \end{bmatrix}$

i. The matrix found in part (h) can now be used to find the coordinates of $A\mathbf{v}$ relative to the basis C. That is

$$\left[A\begin{bmatrix} x \\ y \end{bmatrix}\right]_C = \begin{bmatrix} -1 & 2 \\ 5 & -4 \\ 7 & -5 \\ -2 & 4 \end{bmatrix}\left[\begin{bmatrix} x \\ y \end{bmatrix}\right]_B = \begin{bmatrix} -1 & 2 \\ 5 & -4 \\ 7 & -5 \\ -2 & 4 \end{bmatrix}\begin{bmatrix} \frac{1}{3}x + \frac{1}{3}y \\ \frac{2}{3}x - \frac{1}{3}y \end{bmatrix} = \begin{bmatrix} x-y \\ -x+3y \\ -x+4y \\ 2x-2y \end{bmatrix}.$$

Then

$$A\begin{bmatrix} x \\ y \end{bmatrix} = (x-y)\begin{bmatrix} 1 \\ 0 \\ 1 \\ 1 \end{bmatrix} + (-x+3y)\begin{bmatrix} -1 \\ 1 \\ 0 \\ 1 \end{bmatrix} + (-x+4y)\begin{bmatrix} 1 \\ 0 \\ 0 \\ 0 \end{bmatrix} + (2x-2y)\begin{bmatrix} 0 \\ 1 \\ 0 \\ 0 \end{bmatrix} = \begin{bmatrix} x \\ x+y \\ x-y \\ 2y \end{bmatrix},$$

which agrees with the definition for T found in part (b).

3. a. A reflection through the x-axis is given by the operator $S\begin{bmatrix} x \\ y \end{bmatrix} = \begin{bmatrix} x \\ -y \end{bmatrix}$ and a reflection through the

y-axis by $T\begin{bmatrix} x \\ y \end{bmatrix} = \begin{bmatrix} -x \\ y \end{bmatrix}$. The operator S is a linear operator since

$$S\left(\begin{bmatrix} x \\ y \end{bmatrix} + c\begin{bmatrix} u \\ v \end{bmatrix}\right) = S\left(\begin{bmatrix} x+u \\ y+cv \end{bmatrix}\right) = \begin{bmatrix} x+u \\ -(y+cv) \end{bmatrix} = \begin{bmatrix} x \\ -y \end{bmatrix} + c\begin{bmatrix} u \\ -v \end{bmatrix} = S\left(\begin{bmatrix} x \\ y \end{bmatrix}\right) + cS\left(\begin{bmatrix} u \\ v \end{bmatrix}\right).$$

Similarly, T is also a linear operator.

b. $[S]_B = \begin{bmatrix} 1 & 0 \\ 0 & -1 \end{bmatrix}$ and $[T]_B = \begin{bmatrix} -1 & 0 \\ 0 & 1 \end{bmatrix}$.   **c.** Since $[T \circ S]_B = \begin{bmatrix} -1 & 0 \\ 0 & -1 \end{bmatrix} = [S \circ T]_B$, the linear operators $S \circ T$ and $T \circ S$ reflect a vector through the origin.

5. a. Since $T(\mathbf{v_1}) = \mathbf{v_2} = (0)\mathbf{v_1} + (1)\mathbf{v_2}$ and $T(\mathbf{v_2}) = \mathbf{v_1} = (1)\mathbf{v_1} + (0)\mathbf{v_2}$, then $[T]_B = \begin{bmatrix} 0 & 1 \\ 1 & 0 \end{bmatrix}$

b. Simply switch the column vectors of the matrix found in part (a). That is,

$$[T]_B^{B'} = [\ [T(\mathbf{v_1})]_{B'}\ \ [T(\mathbf{v_2})]_{B'}\ ] = [\ [\mathbf{v_2}]_{B'}\ \ [\mathbf{v_1}]_{B'}\ ] = \begin{bmatrix} 1 & 0 \\ 0 & 1 \end{bmatrix}.$$

7.   a.   The normal vector for the plane is the cross product of the linearly independent vectors $\begin{bmatrix} 1 \\ 0 \\ 0 \end{bmatrix}$ and $\begin{bmatrix} 0 \\ 1 \\ 1 \end{bmatrix}$, that is, $\mathbf{n} = \begin{bmatrix} 0 \\ -1 \\ 1 \end{bmatrix}$. Then using the formula given for the reflection of a vector across a plane with normal $\mathbf{n}$ and the fact that B is the standard basis, we have that $[T]_B =$

$$\left[T\left(\begin{bmatrix} 1 \\ 0 \\ 0 \end{bmatrix}\right)\ \ T\left(\begin{bmatrix} 0 \\ 1 \\ 0 \end{bmatrix}\right)\ \ T\left(\begin{bmatrix} 0 \\ 0 \\ 1 \end{bmatrix}\right) \right] = \begin{bmatrix} 1 & 0 & 0 \\ 0 & 0 & 1 \\ 0 & 1 & 0 \end{bmatrix}.$$

b. $T\begin{bmatrix} -1 \\ 2 \\ 1 \end{bmatrix} = \left[T\begin{bmatrix} -1 \\ 2 \\ 1 \end{bmatrix} \right]_B = \begin{bmatrix} 1 & 0 & 0 \\ 0 & 0 & 1 \\ 0 & 1 & 0 \end{bmatrix}\begin{bmatrix} -1 \\ 2 \\ 1 \end{bmatrix} = \begin{bmatrix} -1 \\ 1 \\ 2 \end{bmatrix}$

c. $N(T) = \left\{ \begin{bmatrix} 0 \\ 0 \\ 0 \end{bmatrix} \right\}$ **d.** $R(T) = \mathbb{R}^3$ **e.** $[T^n]_B = \begin{bmatrix} 1 & 0 & 0 \\ 0 & 0 & 1 \\ 0 & 1 & 0 \end{bmatrix}^n = \begin{cases} I, & \text{if } n \text{ is even} \\ \begin{bmatrix} 1 & 0 & 0 \\ 0 & 0 & 1 \\ 0 & 1 & 0 \end{bmatrix}, & \text{if } n \text{ is odd} \end{cases}$

9. Since $T^2 - T + I = \mathbf{0}, T - T^2 = I$. Then

$$(T \circ (I-T))(\mathbf{v}) = T((I-T)(\mathbf{v})) = T(\mathbf{v} - T(\mathbf{v})) = T(\mathbf{v}) - T^2(\mathbf{v}) = I(\mathbf{v}) = \mathbf{v}.$$

Chapter Test Chapter 4

1. F.

$$T\left(\mathbf{u}+\mathbf{v}\right) \neq T\left(\mathbf{u}\right)+T\left(\mathbf{v}\right)$$

since the second component of the sum will contain a plus 4.

2. F. Since

$$T(x+y) = 2x + 2y - 1$$

but

$$T(x) + T(y) = 2x + 2y - 2.$$

3. T

4. T

5. T

6. F. Since

$$T(\mathbf{u}) = \frac{1}{3}T(2\mathbf{u}-\mathbf{v}) + \frac{1}{3}T(\mathbf{u}+\mathbf{v})$$

$$= \frac{1}{3}\begin{bmatrix} 1 \\ 1 \end{bmatrix} + \frac{1}{3}\begin{bmatrix} 0 \\ 1 \end{bmatrix} = \begin{bmatrix} 1/3 \\ 2/3 \end{bmatrix}.$$

7. F. Since

$$N(T) = \left\{ \begin{bmatrix} 2t \\ t \end{bmatrix} \Bigg| \; t \in \mathbb{R} \right\}$$

8. F. If T is one-to-one, then the set is linearly independent.

9. T

10. F. For example, $T(1) = 0 = T(2)$.

11. T

12. F. Since, for every k,

$$T\left(\begin{bmatrix} k \\ k \end{bmatrix} \right) = \begin{bmatrix} 0 \\ 0 \end{bmatrix}.$$

13. T

14. T

15. T

16. T

17. T

18. T

19. T

20. T

21. F. The transformation is a constant mapping.

22. F. Since $\dim(N(T)) + \dim(R(T)) = \dim(\mathbb{R}^4) = 4$, then $\dim(R(t)) = 2$.

23. T

24. F. Since

$$[T]_B = \begin{bmatrix} 2 & -1 & 1 \\ 1 & 0 & 0 \\ -1 & 1 & 0 \end{bmatrix},$$

then

$$([T]_B)^{-1} = \begin{bmatrix} 0 & 1 & 0 \\ 0 & 1 & 1 \\ 1 & -1 & -1 \end{bmatrix}.$$

25. T

26. F. If T is a linear transformation, then $T(\mathbf{0}) = \mathbf{0}$.

27. F. It projects each vector onto the xz-plane.

28. T

29. F. Define $T : \mathbb{R}^n \to \mathbb{R}$ such that $T(\mathbf{e}_i) = i$ for $i = 1, \ldots, n$ so that $\{T(\mathbf{e}_1), \ldots, T(\mathbf{e}_n)\} = \{1, 2, \ldots, n\}$. This set is not a basis for $\mathbb{R}$, but T is onto.

30. T

31. F. False, let $T : \mathbb{R}^2 \to \mathbb{R}^2$ by $T(\mathbf{v}) = \mathbf{v}$. If B is the standard basis and $B' = \{\mathbf{e_2}, \mathbf{e_2}\}$, then $[T]_B = \begin{bmatrix} 0 & 1 \\ 1 & 0 \end{bmatrix}$.

32. F. Since $N(T)$ consists of only the zero vector, the null space has dimension 0.

33. T

34. F. Any idempotent matrix is in the null space.

35. F. Since $T(p(x))$ has degree at most 2.

36. T.

37. T

38. T

39. F. Let $A = \begin{bmatrix} 1 & 0 \\ 0 & 1 \\ 0 & 0 \end{bmatrix}$, so $T(\mathbf{v}) = A \begin{bmatrix} x \\ y \end{bmatrix} = \begin{bmatrix} x \\ y \\ 0 \end{bmatrix}$ is one-to-one.

40. T.

5 Eigenvalues and Eigenvectors

Exercise Set 5.1

An eigenvalue of the $n \times n$ matrix A is a number λ such that there is a nonzero vector $\mathbf{v}$ with $A\mathbf{v} = \lambda\mathbf{v}$. So if λ and $\mathbf{v}$ are an eigenvalue–eigenvector pair, then the action on $\mathbf{v}$ is a scaling of the vector. Notice that if $\mathbf{v}$ is an eigenvalue corresponding to the eigenvalue λ, then

$$A(c\mathbf{v}) = cA\mathbf{v} = c(\lambda\mathbf{v}) = \lambda(c\mathbf{v}),$$

so A will have infinitely many eigenvectors corresponding to the eigenvalue λ. Also recall that an eigenvalue can be 0 (or a complex number), but eigenvectors are only nonzero vectors. An eigenspace is the set of all eigenvectors corresponding to an eigenvalue λ along with the zero vector, and is denoted by $V_\lambda = \{\mathbf{v} \in \mathbb{R}^n \mid A\mathbf{v} = \lambda\mathbf{v}\}$. Adding the zero vector makes V_λ a subspace. The eigenspace can also be viewed as the null space of $A - \lambda I$. To determine the eigenvalues of a matrix A we have:

$$\lambda \text{ is an eigenvalue of } A \Leftrightarrow \det(A - \lambda I) = 0.$$

The last equation is the characteristic equation for A. As an immediate consequence, if A is a triangular matrix, then the eigenvalues are the entries on the diagonal. To then find the corresponding eigenvectors, for each eigenvalue λ, the equation $A\mathbf{v} = \lambda\mathbf{v}$ is solved for $\mathbf{v}$. An outline of the typical computations for the matrix $A = \begin{bmatrix} -1 & 1 & -2 \\ 1 & -1 & 2 \\ 1 & 0 & 1 \end{bmatrix}$ are:

- To find the eigenvalues solve the equation $\det(A - \lambda I) = 0$. Expanding across row three, we have that

$$\begin{vmatrix} -1-\lambda & 1 & -2 \\ 1 & -1-\lambda & 2 \\ 1 & 0 & 1-\lambda \end{vmatrix} = \begin{vmatrix} 1 & -2 \\ -1-\lambda & 2 \end{vmatrix} + (1-\lambda)\begin{vmatrix} -1-\lambda & 1 \\ 1 & -1-\lambda \end{vmatrix} = -\lambda^2 - \lambda^3.$$

  Then $\det(A - \lambda I) = 0$ if and only if

$$-\lambda^2 - \lambda^3 = -\lambda^2(1+\lambda) = 0, \text{ so the eigenvalues are } \lambda_1 = 0, \lambda_2 = 0.$$

- To find the eigenvectors corresponding to $\lambda_1 = 0$ solve $A\mathbf{v} = \mathbf{0}$. Since

$$\begin{bmatrix} -1 & 1 & -2 \\ 1 & -1 & 2 \\ 1 & 0 & 1 \end{bmatrix} \xrightarrow{\text{reduces to}} \begin{bmatrix} -1 & 1 & -2 \\ 0 & 1 & -1 \\ 0 & 0 & 0 \end{bmatrix},$$

  the eigenvectors are of the form $\begin{bmatrix} -t \\ t \\ t \end{bmatrix}$, for any $t \neq 0$. Similarly, the eigenvectors of $\lambda_2 = -1$ have the from $\begin{bmatrix} -2t \\ 2t \\ t \end{bmatrix}, t \neq 0$.

- The eigenspaces are $V_0 = \left\{ t\begin{bmatrix} -1 \\ 1 \\ 1 \end{bmatrix} \middle| t \in \mathbb{R}^3 \right\}$ and $V_{-1} = \left\{ t\begin{bmatrix} -2 \\ 2 \\ 1 \end{bmatrix} \middle| t \in \mathbb{R}^3 \right\}$.

- Notice that there are only two linearly independent eigenvectors of A, the algebraic multiplicity of $\lambda_1 = 0$ is 2, the algebraic multiplicity of $\lambda_2 = -1$ is 1, and the geometric multiplicities are both 1. For a 3×3 matrix other possibilities are:

◆ The matrix can have three distinct real eigenvalues and three linearly independent eigenvectors.

◆ The matrix can have two distinct real eigenvalues such that one has algebraic multiplicity 2 and the other algebraic multiplicity 1. But the matrix can still have three linearly independent eigenvectors, two from the eigenvalue of multiplicity 2 and one form the other.

■ Solutions to Odd Exercises

1. Since the matrix equation $\begin{bmatrix} 3 & 0 \\ 1 & 3 \end{bmatrix} \begin{bmatrix} 0 \\ 1 \end{bmatrix} = \lambda \begin{bmatrix} 0 \\ 1 \end{bmatrix}$ is satisfied if and only if $\lambda = 3$, the eigenvalue corresponding to the eigenvector $\begin{bmatrix} 0 \\ 1 \end{bmatrix}$ is $\lambda = 3$.

3. The corresponding eigenvalue is $\lambda = 0$.   **5.** The corresponding eigenvalue is $\lambda = 1$.

7. a. The characteristic equation is $\det(A - \lambda I) = 0$, that is,

$$\det(A - \lambda I) = \begin{vmatrix} -2 - \lambda & 2 \\ 3 & -3 - \lambda \end{vmatrix} = (-2 - \lambda)(-3 - \lambda) - 6 = \lambda^2 + 5\lambda = 0.$$

b. Since the eigenvalues are the solutions to the characteristic equation $\lambda^2 + 5\lambda = \lambda(\lambda + 5) = 0$, the eigenvalues are $\lambda_1 = 0$ and $\lambda_2 = -5$. **c.** The corresponding eigenvectors are found by solving, respectively, $\begin{bmatrix} -2 & 2 \\ 3 & -3 \end{bmatrix} \mathbf{v} = 0$ and $\begin{bmatrix} -2 & 2 \\ 3 & -3 \end{bmatrix} \mathbf{v} = -5\mathbf{v}$. Hence the eigenvectors are $\mathbf{v_1} = \begin{bmatrix} 1 \\ 1 \end{bmatrix}$ and $\mathbf{v_2} = \begin{bmatrix} -2 \\ 3 \end{bmatrix}$, respectively.

d.

$$A\mathbf{v_1} = \begin{bmatrix} -2 & 2 \\ 3 & -3 \end{bmatrix} \begin{bmatrix} 1 \\ 1 \end{bmatrix} = \begin{bmatrix} 0 \\ 0 \end{bmatrix} = 0 \begin{bmatrix} 1 \\ 1 \end{bmatrix}, \text{ and } A\mathbf{v_2} = \begin{bmatrix} -2 & 2 \\ 3 & -3 \end{bmatrix} \begin{bmatrix} -2 \\ 3 \end{bmatrix} = \begin{bmatrix} 10 \\ -15 \end{bmatrix} = (-5) \begin{bmatrix} -2 \\ 3 \end{bmatrix}$$

9. a. $(\lambda - 1)^2 = 0$ **b.** $\lambda_1 = 1$ **c.** $\mathbf{v_1} = \begin{bmatrix} 1 \\ 0 \end{bmatrix}$ **d.** $\begin{bmatrix} 1 & -2 \\ 0 & 1 \end{bmatrix} \begin{bmatrix} 1 \\ 0 \end{bmatrix} = \begin{bmatrix} 1 \\ 0 \end{bmatrix} = (1) \begin{bmatrix} 1 \\ 0 \end{bmatrix}$

11. a. The characteristic equation $\det(A - \lambda I) = 0$, is

$$\begin{vmatrix} -1 - \lambda & 0 & 1 \\ 0 & 1 - \lambda & 0 \\ 0 & 2 & -1 - \lambda \end{vmatrix} = 0.$$

Expanding down column one, we have that

$$0 = \begin{vmatrix} -1 - \lambda & 0 & 1 \\ 0 & 1 - \lambda & 0 \\ 0 & 2 & -1 - \lambda \end{vmatrix} = (-1 - \lambda) \begin{vmatrix} 1 - \lambda & 0 \\ 2 & -1 - \lambda \end{vmatrix} = (1 + \lambda)^2 (1 - \lambda).$$

b. $\lambda_1 = -1, \lambda_2 = 1$ **c.** $\mathbf{v_1} = \begin{bmatrix} 1 \\ 0 \\ 0 \end{bmatrix}, \mathbf{v_2} = \begin{bmatrix} 1 \\ 2 \\ 2 \end{bmatrix}$

d.

$$\begin{bmatrix} -1 & 0 & 1 \\ 0 & 1 & 0 \\ 0 & 2 & -1 \end{bmatrix} \begin{bmatrix} 1 \\ 0 \\ 0 \end{bmatrix} = \begin{bmatrix} -1 \\ 0 \\ 0 \end{bmatrix} = (-1) \begin{bmatrix} 1 \\ 0 \\ 0 \end{bmatrix} \text{ and } \begin{bmatrix} -1 & 0 & 1 \\ 0 & 1 & 0 \\ 0 & 2 & -1 \end{bmatrix} \begin{bmatrix} 1 \\ 2 \\ 2 \end{bmatrix} = \begin{bmatrix} 1 \\ 2 \\ 2 \end{bmatrix} = (1) \begin{bmatrix} 1 \\ 2 \\ 2 \end{bmatrix}$$

13. a. $(\lambda - 2)(\lambda - 1)^2 = 0$ **b.** $\lambda_1 = 2, \lambda_2 = 1$ **c.** $\mathbf{v_1} = \begin{bmatrix} 1 \\ 0 \\ 0 \end{bmatrix}, \mathbf{v_2} = \begin{bmatrix} -3 \\ 1 \\ 1 \end{bmatrix}$

d. $\begin{bmatrix} 2 & 1 & 2 \\ 0 & 2 & -1 \\ 0 & 1 & 0 \end{bmatrix} \begin{bmatrix} 1 \\ 0 \\ 0 \end{bmatrix} = \begin{bmatrix} 2 \\ 0 \\ 0 \end{bmatrix} = (2) \begin{bmatrix} 1 \\ 0 \\ 0 \end{bmatrix}$ and $\begin{bmatrix} 2 & 1 & 2 \\ 0 & 2 & -1 \\ 0 & 1 & 0 \end{bmatrix} \begin{bmatrix} -3 \\ 1 \\ 1 \end{bmatrix} = \begin{bmatrix} -3 \\ 1 \\ 1 \end{bmatrix} = (1) \begin{bmatrix} -3 \\ 1 \\ 1 \end{bmatrix}$

15. a. $(\lambda+1)(\lambda-2)(\lambda+2)(\lambda-4) = 0$ **b.** $\lambda_1 = -1, \lambda_2 = 2, \lambda_3 = -2, \lambda_4 = 4$

c. $\mathbf{v_1} = \begin{bmatrix} 1 \\ 0 \\ 0 \\ 0 \end{bmatrix}$, $\mathbf{v_2} = \begin{bmatrix} 0 \\ 1 \\ 0 \\ 0 \end{bmatrix}$, $\mathbf{v_3} = \begin{bmatrix} 0 \\ 0 \\ 1 \\ 0 \end{bmatrix}$, $\mathbf{v_4} = \begin{bmatrix} 0 \\ 0 \\ 0 \\ 1 \end{bmatrix}$ **d.** For the eigenvalue $\lambda = -1$, the verification is

$\begin{bmatrix} -1 & 0 & 0 & 0 \\ 0 & 2 & 0 & 0 \\ 0 & 0 & -2 & 0 \\ 0 & 0 & 0 & 4 \end{bmatrix} \begin{bmatrix} 1 \\ 0 \\ 0 \\ 0 \end{bmatrix} = \begin{bmatrix} -1 \\ 0 \\ 0 \\ 0 \end{bmatrix} = (-1) \begin{bmatrix} 1 \\ 0 \\ 0 \\ 0 \end{bmatrix}$. The other cases are similar.

17. Let $A = \begin{bmatrix} x & y \\ z & w \end{bmatrix}$ and assume the characteristic equation is given by $\lambda^2 + b\lambda + c$. The characteristic equation $\det(A - \lambda I) = 0$, is given by $(x - \lambda)(w - \lambda) - zy = 0$ and simplifies to $\lambda^2 - (x + w)\lambda + (xw - yz) = 0$. Hence, $b = -(x + w) = -\mathbf{tr}(A)$ and $c = xw - yz = \det(A)$.

19. Suppose A is not invertible. Then the homogeneous equation $A\mathbf{x} = \mathbf{0}$ has a nontrivial solution $\mathbf{x_0}$. Since $A\mathbf{x_0} = \mathbf{0} = 0\mathbf{x_0}$, then $\mathbf{x_0}$ is an eigenvector of A corresponding to the eigenvalue $\lambda = 0$. Conversely, suppose that $\lambda = 0$ is an eigenvalue of A. Then there exists a nonzero vector $\mathbf{x_0}$ such that $A\mathbf{x_0} = \mathbf{0}$, so A is not invertible.

21. Let A be an idempotent matrix, that is, A satisfies $A^2 = A$. Also, let λ be an eigenvalue of A with corresponding eigenvector $\mathbf{v}$, so that $A\mathbf{v} = \lambda\mathbf{v}$. We also have that $A\mathbf{v} = A^2\mathbf{v} = A(A\mathbf{v}) = A(\lambda\mathbf{v}) = \lambda A\mathbf{v} = \lambda^2\mathbf{v}$. The two equations $A\mathbf{v} = \lambda\mathbf{v}$ and $A\mathbf{v} = \lambda^2\mathbf{v}$ give $\lambda^2\mathbf{v} = \lambda\mathbf{v}$, so that $(\lambda^2 - \lambda)\mathbf{v} = \mathbf{0}$. But $\mathbf{v}$ is an eigenvector, so that $\mathbf{v} \neq \mathbf{0}$. Hence, $\lambda(\lambda - 1) = 0$, so that the only eigenvalues are either $\lambda = 0$ or $\lambda = 1$.

23. Let A be such that $A^n = \mathbf{0}$ for some n and let λ be an eigenvalue of A with corresponding eigenvector $\mathbf{v}$, so that $A\mathbf{v} = \lambda\mathbf{v}$. Then $A^2\mathbf{v} = \lambda A\mathbf{v} = \lambda^2\mathbf{v}$. Continuing in this way we see that $A^n\mathbf{v} = \lambda^n\mathbf{v}$. Since $A^n = \mathbf{0}$, then $\lambda^n\mathbf{v} = \mathbf{0}$. Since $\mathbf{v} \neq \mathbf{0}$, then $\lambda^n = 0$ and hence, $\lambda = 0$.

25. Since A is invertible, the inverse A^{-1} exists. Notice that for a matrix C, we can use the multiplicative property of the determinant to show that

$$\det(A^{-1}CA) = \det(A^{-1})\det(C)\det(A) = \det(A^{-1})\det(A)\det(C) = \det(A^{-1}A)\det(C) = \det(I)\det(C) = \det(C).$$

Then $\det(AB - \lambda I) = \det(A^{-1}(AB - \lambda I)A) = \det(BA - \lambda I)$, so that AB and BA have the same characteristic equation and hence, the same eigenvalues.

27. Since the matrix is triangular, then the characteristic equation is given by

$$\det(A - \lambda I) = (\lambda - a_{11})(\lambda - a_{22}) \cdots (\lambda - a_{nn}) = 0,$$

so that the eigenvalues are the diagonal entries.

29. Let λ be an eigenvalue of $C = B^{-1}AB$ with corresponding eigenvector $\mathbf{v}$. Since $C\mathbf{v} = \lambda\mathbf{v}$, then $B^{-1}AB\mathbf{v} = \lambda\mathbf{v}$. Multiplying both sides of the previous equation on the left by B gives $A(B\mathbf{v}) = \lambda(B\mathbf{v})$. Therefore, $B\mathbf{v}$ is an eigenvector of A corresponding to λ.

31. The operator that reflects a vector across the x-axis is $T\left(\begin{bmatrix} x \\ y \end{bmatrix}\right) = \begin{bmatrix} x \\ -y \end{bmatrix}$. Then $T\left(\begin{bmatrix} x \\ y \end{bmatrix}\right) = \lambda \begin{bmatrix} x \\ y \end{bmatrix}$ if and only if $x(\lambda - 1) = 0$ and $y(\lambda + 1) = 0$. If $\lambda = 1$, then $y = 0$ and if $\lambda = -1$, then $x = 0$. Hence, the eigenvalues are $\lambda = 1$ and $\lambda = -1$ with corresponding eigenvectors $\begin{bmatrix} 1 \\ 0 \end{bmatrix}$ and $\begin{bmatrix} 0 \\ 1 \end{bmatrix}$, respectively.

33. If $\theta \neq 0$ or $\theta \neq \pi$, then T can only be described as a rotation. Hence $T\left(\begin{bmatrix} x \\ y \end{bmatrix}\right)$ cannot be expressed by scalar multiplication as this only performs a contraction or a dilation. When $\theta = 0$, then T is the identity map $T\left(\begin{bmatrix} x \\ y \end{bmatrix}\right) = \begin{bmatrix} x \\ y \end{bmatrix}$. In this case every nonzero vector in $\mathbb{R}^2$ is an eigenvector with corresponding eigenvalue equal to 1. If $\theta = \pi$, then $T\left(\begin{bmatrix} x \\ y \end{bmatrix}\right) = \begin{bmatrix} -1 & 0 \\ 0 & -1 \end{bmatrix}\begin{bmatrix} x \\ y \end{bmatrix} = -\begin{bmatrix} x \\ y \end{bmatrix}$. In this case every nonzero vector in $\mathbb{R}^2$ is an eigenvector with eigenvalue equal to -1.

35. a. Notice that, for example, $[T(x-1)]_B = [-x^2 - x]_B$ and since $-\frac{1}{2}(x-1) - \frac{1}{2}(x+1) - x^2 = -x - x^2$, then $[-x^2 - x]_B = \begin{bmatrix} -\frac{1}{2} \\ -\frac{1}{2} \\ -1 \end{bmatrix}$. After computing $[T(x+1)]_B$ and $[T(x^2)]_B$ we see that the matrix representation of the operator is given by

$$[T]_B = \begin{bmatrix} [T(x-1)]_B & [T(x+1)]_B & [T(x^2)]_B \end{bmatrix} = \begin{bmatrix} [-x^2 - x]_B & [-x^2 + x]_B & [x^2]_B \end{bmatrix} = \begin{bmatrix} -1/2 & 1/2 & 0 \\ -1/2 & 1/2 & 0 \\ -1 & -1 & 1 \end{bmatrix}.$$

b. Similar to part (a), $[T]_{B'} = \begin{bmatrix} 1 & 1 & 0 \\ -1 & -1 & 0 \\ -1 & 0 & 1 \end{bmatrix}$.

c. Since $x^3 - x^2$ is the characteristic polynomial for both the matrices in part (a) and in (b), the eigenvalues for the two matrices are the same.

Exercise Set 5.2

Diagonalization of a matrix is another type of factorization. A matrix A is diagonalizable if there is an invertible matrix P and a diagonal matrix D such that $A = PDP^{-1}$. An $n \times n$ matrix A is diagonalizable if and only if A has n linearly independent eigenvectors. In this case there is also a process for finding the matrices P and D :

- The column vectors of P are the eigenvectors of A.

- The diagonal entries of D are the corresponding eigenvalues, placed on the diagonal of D in the same order as the eigenvectors in P. If an eigenvalue has multiplicity greater than 1, then it is repeated that many times on the diagonal.

In the following table, three typical examples are given to describe the connections between the algebraic multiplicity of an eigenvalue, the geometric multiplicity of an eigenvalue, and whether or not a matrix is diagonalizable.

Three Examples

$$A = \begin{bmatrix} 0 & -1 & 2 \\ 0 & -1 & 2 \\ 1 & 1 & -1 \end{bmatrix}$$

Eigenvalues:
$-3, 0, 1$
Eigenvectors:

$$\begin{bmatrix} 1 \\ 1 \\ -1 \end{bmatrix}, \begin{bmatrix} -1 \\ 2 \\ 1 \end{bmatrix}, \begin{bmatrix} 1 \\ 1 \\ 1 \end{bmatrix}$$

Diagonalizable:

$$P = \begin{bmatrix} 1 & -1 & 1 \\ 1 & 2 & 1 \\ -1 & 1 & 1 \end{bmatrix}$$

$$D = \begin{bmatrix} -3 & 0 & 0 \\ 0 & 0 & 0 \\ 0 & 0 & 1 \end{bmatrix}$$

$$A = \begin{bmatrix} -1 & 2 & 0 \\ -1 & 2 & 0 \\ 1 & -1 & 1 \end{bmatrix}$$

Eigenvalues:
$0, 1$ multiplicity 2
Eigenvectors:

$$\begin{bmatrix} 2 \\ 1 \\ -1 \end{bmatrix}, \begin{bmatrix} 1 \\ 1 \\ 0 \end{bmatrix}, \begin{bmatrix} 0 \\ 0 \\ 1 \end{bmatrix}$$

Diagonalizable:

$$P = \begin{bmatrix} 2 & 1 & 0 \\ 1 & 1 & 0 \\ -1 & 0 & 1 \end{bmatrix}$$

$$D = \begin{bmatrix} 0 & 0 & 0 \\ 0 & 1 & 0 \\ 0 & 0 & 1 \end{bmatrix}$$

$$A = \begin{bmatrix} -1 & 0 & 0 \\ 2 & 0 & 0 \\ -1 & 1 & 0 \end{bmatrix}$$

Eigenvalues:
0 multiplicity 2, -1
Eigenvectors:

$$\begin{bmatrix} 0 \\ 0 \\ 1 \end{bmatrix}, \begin{bmatrix} 1 \\ -2 \\ 3 \end{bmatrix}$$

Not Diagonalizable.

Other useful results given in Section 5.2 are:

- If an $n \times n$ matrix A has n distinct eigenvalues, then A is diagonalizable.

- Similar matrices have the same eigenvalues.

- If $T : V \longrightarrow V$ is a linear operator and B_1 and B_2 are bases for V, then $[T]_{B_1}$ and $[T]_{B_2}$ have the same eigenvalues.

- An $n \times n$ matrix A is diagonalizable if and only if the algebraic and geometric multiplicities add up correctly. That is, suppose A has eigenvalues $\lambda_1, \ldots, \lambda_k$ with algebraic multiplicities $d_1, \ldots, d_k$. The matrix A is diagonalizable if and only if

$$d_1 + d_2 + \cdots + d_k = \dim(V_{\lambda_1}) + \dim(V_{\lambda_2}) + \cdots + \dim(V_{\lambda_k}) = n.$$

■ Solutions to Odd Exercises

1. $P^{-1}AP = \begin{bmatrix} 1 & 0 \\ 0 & -3 \end{bmatrix}$

3. $P^{-1}AP = \begin{bmatrix} 0 & 0 & 0 \\ 0 & -2 & 0 \\ 0 & 0 & 1 \end{bmatrix}$

5. The eigenvalues for the matrix A are -2 and -1. The eigenvectors are not necessary in this case, since A is a 2×2 matrix with two distinct eigenvalues. Hence, A is diagonalizable.

7. There is only one eigenvalue -1, with multiplicity 2 and eigenvector $\begin{bmatrix} 1 \\ 0 \end{bmatrix}$. Since A does not have two linearly independent eigenvectors, A is not diagonalizable.

9. The eigenvalues for the matrix A are 1 and 0. The eigenvectors are not necessary in this case, since A is a 2×2 matrix with two distinct eigenvalues. Hence, A is diagonalizable.

11. The eigenvalues for the matrix A are $3, 4$, and 0. The eigenvectors are not necessary in this case, since A is a 3×3 matrix with three distinct eigenvalues. Hence, A is diagonalizable.

13. The eigenvalues are -1, and 2 with multiplicity 2, and corresponding linearly independent eigenvectors $\begin{bmatrix} 1 \\ -5 \\ 2 \end{bmatrix}$ and $\begin{bmatrix} -1 \\ -1 \\ 1 \end{bmatrix}$, respectively. Since there are only two linearly independent eigenvectors, A is not diagonalizable

15. The matrix A has two eigenvalues 1, and 0 with multiplicity 2. So the eigenvectors are needed to determine if the matrix is diagonalizable. In this case there are three linearly independent eigenvectors $\begin{bmatrix} -1 \\ 1 \\ 1 \end{bmatrix}$, $\begin{bmatrix} 0 \\ 1 \\ 0 \end{bmatrix}$, and $\begin{bmatrix} 0 \\ 0 \\ 1 \end{bmatrix}$. Hence, A is diagonalizable.

17. The matrix A has eigenvalues, $-1, 2$, and 0 with multiplicity 2. In addition there are four linearly independent eigenvectors $\begin{bmatrix} 0 \\ -1 \\ 1 \\ 0 \end{bmatrix}$, $\begin{bmatrix} 0 \\ 1 \\ 2 \\ 3 \end{bmatrix}$, $\begin{bmatrix} 0 \\ -1 \\ 0 \\ 1 \end{bmatrix}$, and $\begin{bmatrix} -1 \\ 0 \\ 1 \\ 0 \end{bmatrix}$ corresponding to the two distinct eigenvalues. Hence, A is diagonalizable.

19. To diagonalize the matrix we first find the eigenvalues. Since the characteristic equation is $\begin{vmatrix} 2-\lambda & 0 \\ -1 & -1-\lambda \end{vmatrix} = (2-\lambda)(-1-\lambda) = 0$, the eigenvalues are $\lambda_1 = 2$ and $\lambda_2 = -1$. The corresponding eigenvectors are found by solving $A \begin{bmatrix} x \\ y \end{bmatrix} = 2 \begin{bmatrix} x \\ y \end{bmatrix}$ and $A \begin{bmatrix} x \\ y \end{bmatrix} = - \begin{bmatrix} x \\ y \end{bmatrix}$. This gives eigenvectors $\begin{bmatrix} -3 \\ 1 \end{bmatrix}$ corresponding to $\lambda_1 = 2$ and $\begin{bmatrix} 0 \\ 1 \end{bmatrix}$ corresponding to $\lambda_2 = -1$. The matrix A is diagonalizable since there are two distinct eigenvalues (or two linearly independent eigenvectors) and if $P = \begin{bmatrix} -3 & 0 \\ 1 & 1 \end{bmatrix}$, then $P^{-1}AP = \begin{bmatrix} 2 & 0 \\ 0 & -1 \end{bmatrix}$.

21. The eigenvalues of A are $-1, 1$, and 0 with corresponding linearly independent eigenvectors $\begin{bmatrix} 0 \\ 1 \\ 1 \end{bmatrix}$, $\begin{bmatrix} 2 \\ 1 \\ 3 \end{bmatrix}$, and $\begin{bmatrix} 0 \\ 1 \\ 2 \end{bmatrix}$, respectively. If $P = \begin{bmatrix} 0 & 2 & 0 \\ 1 & 1 & 1 \\ 1 & 3 & 2 \end{bmatrix}$, then $P^{-1}AP = \begin{bmatrix} -1 & 0 & 0 \\ 0 & 1 & 0 \\ 0 & 0 & 0 \end{bmatrix}$.

23. The eigenvalues of the matrix A are $\lambda_1 = -1$ and $\lambda_2 = 1$ of multiplicity two. But there are three linearly independent eigenvectors, $\begin{bmatrix} 2 \\ 1 \\ 0 \end{bmatrix}$, corresponding to λ_1 and $\begin{bmatrix} 0 \\ 1 \\ 0 \end{bmatrix}$, $\begin{bmatrix} 0 \\ 0 \\ 1 \end{bmatrix}$ corresponding to λ_2. If $P = \begin{bmatrix} 2 & 0 & 0 \\ 1 & 1 & 0 \\ 0 & 0 & 1 \end{bmatrix}$, then $P^{-1}AP = \begin{bmatrix} -1 & 0 & 0 \\ 0 & 1 & 0 \\ 0 & 0 & 1 \end{bmatrix}$.

25. The matrix A has three eigenvalues $\lambda_1 = 1$ of multiplicity 2, $\lambda_2 = 0$, and $\lambda_3 = 2$. There are four linearly independent eigenvectors corresponding to the eigenvalues given as the column vectors in $P = \begin{bmatrix} -1 & 0 & -1 & 1 \\ 0 & 1 & 0 & 0 \\ 0 & 0 & 1 & 1 \\ 1 & 0 & 0 & 0 \end{bmatrix}$. Then $P^{-1}AP = \begin{bmatrix} 1 & 0 & 0 & 0 \\ 0 & 1 & 0 & 0 \\ 0 & 0 & 0 & 0 \\ 0 & 0 & 0 & 2 \end{bmatrix}$.

27. The proof is by induction on the power k. The base case is $k = 1$, which is upheld since $D = P^{-1}AP$, gives $A = PDP^{-1}$. The inductive hypothesis is to assume the result holds for a natural number k. Now assume that $A^k = PD^kP^{-1}$. We need to show that the result holds for the next positive integer $k+1$. Since $A^{k+1} = A^kA$, by the inductive hypothesis, we have that

$$A^{k+1} = A^kA = (PD^kP^{-1})A = (PD^kP^{-1})(PDP^{-1}) = (PD^k)(P^{-1}P)(DP^{-1}) = PD^{k+1}P^{-1}.$$

29. The eigenvalues of the matrix A are 0 and 1 of multiplicity 2, with corresponding linearly independent eigenvectors the column vectors of $P = \begin{bmatrix} 1 & 0 & 1 \\ 1 & -2 & 2 \\ 1 & 1 & 0 \end{bmatrix}$. If $D = \begin{bmatrix} 0 & 0 & 0 \\ 0 & 1 & 0 \\ 0 & 0 & 1 \end{bmatrix}$ is the diagonal matrix with diagonal entries the eigenvalues of A, then $A = PDP^{-1}$. Notice that the eigenvalues on the diagonal have the same order as the corresponding eigenvectors in P, with the eigenvalue 1 repeated two times since the algebraic multiplicity is 2. Since $D^k = \begin{bmatrix} 0^k & 0 & 0 \\ 0 & 1^k & 0 \\ 0 & 0 & 1^k \end{bmatrix} = D$, then for any $k \geq 1$, $A^k = PD^kP^{-1} = PDP^{-1} = \begin{bmatrix} 3 & -1 & -2 \\ 2 & 0 & -2 \\ 2 & -1 & -1 \end{bmatrix}$.

31. Since A is diagonalizable there is an invertible P and diagonal D such that $A = PDP^{-1}$, so that $D = P^{-1}AP$. Since B is similar to A there is an invertible Q such that $B = Q^{-1}AQ$, so that $A = QBQ^{-1}$. Then

$$D = P^{-1}QBQ^{-1}P = (Q^{-1}P)^{-1}B(Q^{-1}P)$$

and hence, B is diagonalizable.

33. If A is diagonalizable with an eigenvalue of multiplicity n, then $A = P(\lambda I)P^{-1} = (\lambda I)PP^{-1} = \lambda I$. Conversely, if $A = \lambda I$, then A is a diagonal matrix.

35. a. $[T]_{B_1} = \begin{bmatrix} 0 & 1 & 0 \\ 0 & 0 & 2 \\ 0 & 0 & 0 \end{bmatrix}$ **b.** $[T]_{B_2} = \begin{bmatrix} 1 & 1 & 2 \\ -1 & -1 & 0 \\ 0 & 0 & 0 \end{bmatrix}$ **c.** The only eigenvalue of A and B is $\lambda = 0$ of multiplicity 3. **d.** Neither matrix has three linearly independent eigenvectors. For example, for $[T]_{B_2}$, the eigenvector corresponding to $\lambda = 0$ is $\begin{bmatrix} -1 \\ 1 \\ 0 \end{bmatrix}$. Thus, the operator T is not diagonalizable.

37. To show that T is not diagonalizable it suffices to show that $[T]_B$ is not diagonalizable for any basis B of $\mathbb{R}^3$. Let B be the standard basis for $\mathbb{R}^3$, so that

$$[T]_B = [\, T(\mathbf{e_1})\ T(\mathbf{e_2})\ T(\mathbf{e_3}) \,] = \begin{bmatrix} 2 & 2 & 2 \\ -1 & 2 & 1 \\ 1 & -1 & 0 \end{bmatrix}.$$

The eigenvalues of $[T]_B$ are $\lambda_1 = 1$ with multiplicity 2, and $\lambda_2 = 2$. The corresponding eigenvectors are $\begin{bmatrix} 0 \\ -1 \\ 1 \end{bmatrix}, \begin{bmatrix} 1 \\ -1 \\ 1 \end{bmatrix}$, respectively. Since there are only two linearly independent eigenvectors, $[T]_B$ and hence, T is not diagonalizable.

39. Since A and B are matrix representations for the same linear operator, they are similar. Let $A = Q^{-1}BQ$. The matrix A is diagonalizable if and only if $D = P^{-1}AP$ for some invertible matrix P and diagonal matrix D. Then

$$D = P^{-1}(Q^{-1}BQ)P = (QP)^{-1}B(QP),$$

so that B is diagonalizable. The proof of the converse is identical.

Exercise Set 5.3

1. The strategy is to uncouple the system of differential equations. Writing the system in matrix form, we have that

$$\mathbf{y}' = A\mathbf{y} = \begin{bmatrix} -1 & 1 \\ 0 & -2 \end{bmatrix} \mathbf{y}.$$

The next step is to diagonalize the matrix A. Since A is triangular the eigenvalues of A are the diagonal entries -1 and -2, with corresponding eigenvectors $\begin{bmatrix} 1 \\ 0 \end{bmatrix}$ and $\begin{bmatrix} -1 \\ 1 \end{bmatrix}$, respectively. So $A = PDP^{-1}$, where $P = \begin{bmatrix} 1 & -1 \\ 0 & 1 \end{bmatrix}$, $P^{-1} = \begin{bmatrix} 1 & 1 \\ 0 & 1 \end{bmatrix}$, and $D = \begin{bmatrix} -1 & 0 \\ 0 & -2 \end{bmatrix}$. The related uncoupled system is $\mathbf{w}' = P^{-1}AP\mathbf{w} = \begin{bmatrix} -1 & 0 \\ 0 & -2 \end{bmatrix}\mathbf{w}$. The general solution to the uncoupled system is $\mathbf{w}(t) = \begin{bmatrix} e^{-t} & 0 \\ 0 & e^{-2t} \end{bmatrix}\mathbf{w}(0)$.

Finally, the general solution to the original system is given by $\mathbf{y}(t) = P\begin{bmatrix} e^{-t} & 0 \\ 0 & e^{-2t} \end{bmatrix}P^{-1}\mathbf{y}(0)$. That is,

$$y_1(t) = (y_1(0) + y_2(0))e^{-t} - y_2(0)e^{-2t}, \quad y_2(t) = y_2(0)e^{-2t}.$$

3. Using the same approach as in Exercise (1), we let $A = \begin{bmatrix} 1 & -3 \\ -3 & 1 \end{bmatrix}$. The eigenvalues of A are 4 and -2 with corresponding eigenvectors $\begin{bmatrix} -1 \\ 1 \end{bmatrix}$ and $\begin{bmatrix} 1 \\ 1 \end{bmatrix}$, respectively, so that

$A = \frac{1}{2}\begin{bmatrix} -1 & 1 \\ 1 & 1 \end{bmatrix}\begin{bmatrix} 4 & 0 \\ 0 & -2 \end{bmatrix}\begin{bmatrix} -1 & 1 \\ 1 & 1 \end{bmatrix}$. So the general solution is given by $\mathbf{y}(t) = P\begin{bmatrix} e^{4t} & 0 \\ 0 & e^{-2t} \end{bmatrix}P^{-1}\mathbf{y}(0)$, that is

$$y_1(t) = \frac{1}{2}(y_1(0) - y_2(0))e^{4t} + \frac{1}{2}(y_1(0) + y_2(0))e^{-2t}, \quad y_2(t) = \frac{1}{2}(-y_1(0) + y_2(0))e^{4t} + \frac{1}{2}(y_1(0) + y_2(0))e^{-2t}.$$

5. Let $A = \begin{bmatrix} -4 & -3 & -3 \\ 2 & 3 & 3 \\ 4 & 2 & 3 \end{bmatrix}$. The eigenvalues of A are $-1, 1$ and 2, so that $A = PDP^{-1} = \begin{bmatrix} -1 & 0 & 1 \\ 0 & -1 & -2 \\ 1 & 1 & 0 \end{bmatrix}\begin{bmatrix} -1 & 0 & 0 \\ 0 & 1 & 0 \\ 0 & 0 & 2 \end{bmatrix}\begin{bmatrix} -2 & -2 & -1 \\ 2 & 1 & 2 \\ -1 & -1 & -1 \end{bmatrix}$, , where the column vectors of P are the eigenvectors of A and the diagonal entries of D are the corresponding eigenvalues. Then $\mathbf{y}(t) = P\begin{bmatrix} e^{-t} & 0 & 0 \\ 0 & e^{t} & 0 \\ 0 & 0 & e^{2t} \end{bmatrix}P^{-1}\mathbf{y}(0)$, that is

$$y_1(t) = (2y_1(0) + y_2(0) + y_3(0))e^{-t} + (-y_1(0) - y_2(0) - y_3(0))e^{2t}$$
$$y_2(t) = (-2y_1(0) - y_2(0) - 2y_3(0))e^{t} + 2(y_1(0) + y_2(0) + y_3(0))e^{2t}$$
$$y_3(t) = (-2y_1(0) - y_2(0) - y_3(0))e^{-t} + (2y_1(0) + y_2(0) + 2y_3(0))e^{t}.$$

7. First find the general solution and then apply the initial conditions. The eigenvalues of $A = \begin{bmatrix} -1 & 0 \\ 2 & 1 \end{bmatrix}$ are 1 and -1, with corresponding eigenvectors $\begin{bmatrix} 0 \\ 1 \end{bmatrix}$ and $\begin{bmatrix} -1 \\ 1 \end{bmatrix}$, respectively. So the general solution is given by $\mathbf{y}(t) = P\begin{bmatrix} e^{t} & 0 \\ 0 & e^{-t} \end{bmatrix}P^{-1}\mathbf{y}(0)$, where the column vectors of P are the eigenvectors of A. Then $y_1(t) = e^{-t}y_1(0), y_2(t) = e^{t}(y_1(0) + y_2(0)) - e^{-t}y_1(0)$. Since $y_1(0) = 1$ and $y_2(0) = -1$, the solution to the initial value problem is $y_1(t) = e^{-t}, y_2(t) = -e^{-t}$.

9. a. Once we construct a system of differential equations that describe the problem, then the solution is found using the same techniques. Let $y_1(t)$ and $y_2(t)$ denote the amount of salt in each take after t minutes, so that $y_1'(t)$ and $y_2'(t)$ are the rates of change for the amount of salt in each tank. The system satisfies the balance law that the rate of change of salt in the tank is equal to the rate in minus the rate out. This gives the initial value problem

$$y_1'(t) = -\frac{1}{60}y_1 + \frac{1}{120}y_2, \quad y_2'(t) = \frac{1}{60}y_1 - \frac{1}{120}y_2, \quad y_1(0) = 12, y_2(0) = 0.$$

b. The solution to the system is $y_1(t) = 4 + 8e^{-\frac{1}{40}t}$, $y_2(t) = 8 - 8e^{-\frac{1}{40}t}$.

c. Since the exponentials in both $y_1(t)$ and $y_2(t)$ go to 0 as t goes to infinity, then $\lim_{t\to\infty} y_1(t) = 4$ and $\lim_{t\to\infty} y_2(t) = 8$. This makes sense since it says that eventually the 12 pounds of salt will be evenly distributed in a ratio of one to two between the two tanks.

Exercise Set 5.4

1. a. Since 15% of the city residents move to the suburbs, 85% remain in the city and since 8% of the suburban population move to the city, 92% remain in the suburbs, so that the transition matrix is given by

$T = \begin{bmatrix} 0.85 & 0.08 \\ 0.15 & 0.92 \end{bmatrix}$. Notice that the transition matrix is a probability matrix since each column sum is 1.

Also since the total population is 2 million with 1.4 currently in the city, the population of the city is $\frac{1.4}{2} = 0.7$ and the population of the suburbs is $\frac{0.6}{2} = 0.3$. Hence, the initial probability vector describing the current

population is $\begin{bmatrix} 0.7 \\ 0.3 \end{bmatrix}$.

b. After 10 years an estimate for the population distribution is given by $T^{10}\begin{bmatrix} 0.7 \\ 0.3 \end{bmatrix} \approx \begin{bmatrix} 0.37 \\ 0.63 \end{bmatrix}$.

c. The steady state probability vector is $\begin{bmatrix} 0.35 \\ 0.65 \end{bmatrix}$, which is the unit probability eigenvector of T corresponding to the eigenvalue $\lambda = 1$.

3. The transition matrix is given by the data in the table, so $T = \begin{bmatrix} 0.5 & 0.4 & 0.1 \\ 0.4 & 0.4 & 0.2 \\ 0.1 & 0.2 & 0.7 \end{bmatrix}$. Since initially

there are only plants with pink flowers, the initial probability state vector is $\begin{bmatrix} 0 \\ 1 \\ 0 \end{bmatrix}$. After three generations

the probabilities of each variety are given by $T^3 \begin{bmatrix} 0 \\ 1 \\ 0 \end{bmatrix} \approx \begin{bmatrix} 0.36 \\ 0.35 \\ 0.29 \end{bmatrix}$ and after 10 generations $T^{10} \begin{bmatrix} 0 \\ 1 \\ 0 \end{bmatrix} \approx$

$\begin{bmatrix} 0.33 \\ 0.33 \\ 0.33 \end{bmatrix}$.

5. The transition matrix for the disease is $T = \begin{bmatrix} 0.5 & 0 & 0 \\ 0.5 & 0.75 & 0 \\ 0 & 0.25 & 1 \end{bmatrix}$. The eigenvector of T, corresponding to

the eigenvalue $\lambda = 1$ is $\begin{bmatrix} 0 \\ 0 \\ 1 \end{bmatrix}$. Hence, the steady state probability vector is $\begin{bmatrix} 0 \\ 0 \\ 1 \end{bmatrix}$ and the disease will not

be eradicated.

7. a. $T = \begin{bmatrix} 0.33 & 0.25 & 0.17 & 0.25 \\ 0.25 & 0.33 & 0.25 & 0.17 \\ 0.17 & 0.25 & 0.33 & 0.25 \\ 0.25 & 0.17 & 0.25 & 0.33 \end{bmatrix}$ **b.** $T\begin{bmatrix} 1 \\ 0 \\ 0 \\ 0 \end{bmatrix} = \begin{bmatrix} 0.5(0.16)^n + 0.25 \\ 0.25 \\ -0.5(0.16)^n + 0.25 \\ 0.25 \end{bmatrix}$ **c.** $\begin{bmatrix} 0.25 \\ 0.25 \\ 0.25 \\ 0.25 \end{bmatrix}$

9. The eigenvalues of T are $\lambda_1 = -q + p + 1$ and $\lambda_2 = 1$, with corresponding eigenvectors $\begin{bmatrix} -1 \\ 1 \end{bmatrix}$ and $\begin{bmatrix} q/p \\ 1 \end{bmatrix}$. Then the steady state probability vector is

$$\frac{1}{1 + \frac{q}{p}} \begin{bmatrix} q/p \\ 1 \end{bmatrix} = \begin{bmatrix} \frac{q}{p+q} \\ \frac{p}{p+q} \end{bmatrix}.$$

Review Exercises Chapter 5

1. a. Let $A = \begin{bmatrix} a & b \\ b & a \end{bmatrix}$. To show that $\begin{bmatrix} 1 \\ 1 \end{bmatrix}$ is an eigenvector of A we need a constant λ such that $A\begin{bmatrix} 1 \\ 1 \end{bmatrix} = \lambda \begin{bmatrix} 1 \\ 1 \end{bmatrix}$. But $\begin{bmatrix} a & b \\ b & a \end{bmatrix}\begin{bmatrix} 1 \\ 1 \end{bmatrix} = \begin{bmatrix} a+b \\ a+b \end{bmatrix} = (a+b)\begin{bmatrix} 1 \\ 1 \end{bmatrix}$. **b.** Since the characteristic equation for A is

$$\det(A - \lambda I) = \begin{vmatrix} a - \lambda & b \\ b & a - \lambda \end{vmatrix} = (a-\lambda)^2 - b^2 = \lambda^2 - 2a\lambda + (a^2 + b^2) = (\lambda - (a+b))(\lambda - (a-b)) = 0,$$

the eigenvalues of A are $\lambda_1 = a + b$, $\lambda_2 = a - b$. **c.** $\mathbf{v_1} = \begin{bmatrix} 1 \\ 1 \end{bmatrix}$, $\mathbf{v_2} = \begin{bmatrix} -1 \\ 1 \end{bmatrix}$ **d.** First notice that A is diagonalizable since it has two linearly independent eigenvectors. The matrix $D = P^{-1}AP$, where P is the matrix with column vectors the eigenvectors and D is the diagonal matrix with the corresponding eigenvalues on the diagonal. That is, $P = \begin{bmatrix} 1 & -1 \\ 1 & 1 \end{bmatrix}$ and $D = \begin{bmatrix} a+b & 0 \\ 0 & a-b \end{bmatrix}$.

3. a. The eigenvalues are $\lambda_1 = 0$ and $\lambda_2 = 1$. **b.** No conclusion can be made from part (a) about whether or not A is diagonalizable, since the matrix does not have four distinct eigenvalues. **c.** Each eigenvalue has multiplicity 2. The eigenvectors corresponding to $\lambda_1 = 0$ are $\begin{bmatrix} 0 \\ 0 \\ 0 \\ 1 \end{bmatrix}$ and $\begin{bmatrix} -1 \\ 0 \\ 1 \\ 0 \end{bmatrix}$ while the eigenvector corresponding to $\lambda_2 = 1$ is $\begin{bmatrix} 0 \\ 1 \\ 0 \\ 0 \end{bmatrix}$. **d.** The eigenvectors found in part (c) are linearly independent. **e.** Since A is a 4×4 matrix with only three linearly independent eigenvectors, then A is not diagonalizable.

5. a. Expanding the determinant of $A - \lambda I$ across row two gives

$$\det(A - \lambda I) = \begin{vmatrix} -\lambda & 1 & 0 \\ 0 & -\lambda & 1 \\ -k & 3 & -\lambda \end{vmatrix} = -(0)\begin{vmatrix} 1 & 0 \\ -3 & -\lambda \end{vmatrix} + (-\lambda)\begin{vmatrix} -\lambda & 0 \\ -k & -\lambda \end{vmatrix} - (1)\begin{vmatrix} -\lambda & 1 \\ -k & 3 \end{vmatrix} = \lambda^3 - 3\lambda + k.$$

Hence, the characteristic equation is $\lambda^3 - 3\lambda + k = 0$.

b. The graphs of $y(\lambda) = \lambda^3 - 3\lambda + k$ for different values of k are shown in the figure.

c. The matrix will have three distinct real eigenvalues when the graph of $y(\lambda) = \lambda^3 - 3\lambda + k$ crosses the x-axis three time. That is, for
$$-2 < k < 2.$$

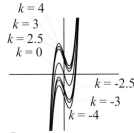

$k = 4$
$k = 3$
$k = 2.5$
$k = 0$
$k = $ -2.5
$k = $ -3
$k = $ -4

7. a. Let $\mathbf{v} = \begin{bmatrix} 1 \\ 1 \\ \vdots \\ 1 \end{bmatrix}$. Then each component of the vector $A\mathbf{v}$ has the same value equal to the common row sum λ. That is, $A\mathbf{v} = \begin{bmatrix} \lambda \\ \lambda \\ \vdots \\ \lambda \end{bmatrix} = \lambda \begin{bmatrix} 1 \\ 1 \\ \vdots \\ 1 \end{bmatrix}$, so λ is an eigenvalue of A corresponding to the eigenvector $\mathbf{v}$. **b.**

Since A and A^t have the same eigenvalues, then the same result holds if the sum of each column of A is equal to λ.

9. a. Suppose $\mathbf{w}$ is in $S(V_{\lambda_0})$, so that $\mathbf{w} = S(\mathbf{v})$ for some eigenvector $\mathbf{v}$ of T corresponding to λ_0. Then

$$T(\mathbf{w}) = T(S(\mathbf{v})) = S(T(\mathbf{v})) = S(\lambda_0\mathbf{v}) = \lambda_0 S(\mathbf{v}) = \lambda_0\mathbf{w}.$$

Hence, $S(V_{\lambda_0}) \subseteq V_{\lambda_0}$.

b. Let $\mathbf{v}$ be an eigenvector of T corresponding to the eigenvalue λ_0. Since T has n distinct eigenvalues then $\dim(V_{\lambda_0}) = 1$ with $V_{\lambda_0} = \mathbf{span}\{\mathbf{v}\}$. Now by part (a), $T(S(\mathbf{v})) = \lambda_0(S(\mathbf{v}))$, so that $S(\mathbf{v})$ is also an eigenvector of T and in $\mathbf{span}\{\mathbf{v}\}$. Consequently, there exists a scalar μ_0 such that $S(\mathbf{v}) = \mu_0\mathbf{v}$, so that $\mathbf{v}$ is also an eigenvector of S.

c. Let $B = \{\mathbf{v_1}, \mathbf{v_2}, \ldots, \mathbf{v_n}\}$ be a basis for V consisting of eigenvectors of T and S. Thus there exist scalars $\lambda_1, \lambda_2, \ldots, \lambda_n$ and $\mu_1, \mu_2, \ldots, \mu_n$ such that $T(\mathbf{v_i}) = \lambda_i\mathbf{v_i}$ and $S(\mathbf{v_i}) = \mu_i\mathbf{v_i}$, for $1 \le i \le n$. Now let $\mathbf{v}$ be a vector in V. Since B is a basis for V then there are scalars $c_1, c_2, \ldots, c_n$ such that $\mathbf{v} = c_1\mathbf{v_1} + c_2\mathbf{v_2} + \ldots + c_n\mathbf{v_n}$. Applying the operator ST to both sides of this equation we obtain

$$ST(\mathbf{v}) = ST(c_1\mathbf{v_1} + c_2\mathbf{v_2} + \ldots + c_n\mathbf{v_n}) = S(c_1\lambda_1\mathbf{v_1} + c_2\lambda_2\mathbf{v_2} + \ldots + c_n\lambda_n\mathbf{v_n})$$
$$= c_1\lambda_1\mu_1\mathbf{v_1} + c_2\lambda_2\mu_2\mathbf{v_2} + \ldots + c_n\lambda_n\mu_n\mathbf{v_n} = c_1\mu_1\lambda_1\mathbf{v_1} + c_2\mu_2\lambda_2\mathbf{v_2} + \ldots + c_n\mu_n\lambda_n\mathbf{v_n}$$
$$= T(c_1\mu_1\mathbf{v_1} + c_2\mu_2\mathbf{v_2} + \ldots + c_n\mu_n\mathbf{v_n}) = TS(c_1\mathbf{v_1} + c_2\mathbf{v_2} + \ldots + c_n\mathbf{v_n}) = TS(\mathbf{v}).$$

Since this holds for all $\mathbf{v}$ in V, then $ST = TS$.

d. The linearly independent vectors $\begin{bmatrix} 1 \\ 0 \\ 1 \end{bmatrix}$, $\begin{bmatrix} -1 \\ 0 \\ 1 \end{bmatrix}$, and $\begin{bmatrix} 0 \\ 1 \\ 0 \end{bmatrix}$ are eigenvectors for both matrices A and B. Let $P = \begin{bmatrix} 1 & -1 & 0 \\ 0 & 0 & 1 \\ 1 & 1 & 0 \end{bmatrix}$. Then

$$P^{-1}AP = \begin{bmatrix} 1/2 & 0 & 1/2 \\ -1/2 & 0 & 1/2 \\ 0 & 1 & 0 \end{bmatrix} \begin{bmatrix} 3 & 0 & 1 \\ 0 & 2 & 0 \\ 1 & 0 & 3 \end{bmatrix} \begin{bmatrix} 1 & -1 & 0 \\ 0 & 0 & 1 \\ 1 & 1 & 0 \end{bmatrix} = \begin{bmatrix} 4 & 0 & 0 \\ 0 & 2 & 0 \\ 0 & 0 & 2 \end{bmatrix}$$

and

$$P^{-1}BP = \begin{bmatrix} 1/2 & 0 & 1/2 \\ -1/2 & 0 & 1/2 \\ 0 & 1 & 0 \end{bmatrix} \begin{bmatrix} 1 & 0 & -2 \\ 0 & 1 & 0 \\ -2 & 0 & 1 \end{bmatrix} \begin{bmatrix} 1 & -1 & 0 \\ 0 & 0 & 1 \\ 1 & 1 & 0 \end{bmatrix} = \begin{bmatrix} -1 & 0 & 0 \\ 0 & 3 & 0 \\ 0 & 0 & 1 \end{bmatrix}.$$

Chapter Test Chapter 5

1. F.

$$P^{-1}AP = \begin{bmatrix} -1 & 2 \\ 0 & -2 \end{bmatrix}$$

2. F. If two matrices are similar, then they have the same eigenvalues. The eigenvalues of A are -1 and 1, and the only eigenvalue of D is -1, so the matrices are not similar.

3. F. The matrix has only two linearly independent eigenvectors $\begin{bmatrix} 2 \\ 0 \\ 1 \end{bmatrix}$ and $\begin{bmatrix} 0 \\ 0 \\ 1 \end{bmatrix}$.

4. T

5. T

6. T

7. T

8. T

9. T

10. T

11. F. Since $\det(A - \lambda I)$ is

$$(\lambda - (a+b))(\lambda - (a-b)),$$

the eigenvalues are $a+b$ and $a-b$.

12. F. The $\det(A - \lambda I)$ is

$$\lambda^2 - 2\lambda + (1 - k),$$

so that, for example, if $k = 1$, then A has eigenvalues 0 and 2.

13. F. Let $A = \begin{bmatrix} 2 & 1 \\ 0 & 2 \end{bmatrix}$.

14. T

15. F. The matrix has one eigenvalue $\lambda = 1$, of multiplicity 2, but does not have two linearly independent eigenvectors.

16. T

17. T

18. T

19. T

20. F.

$$\begin{bmatrix} 2\alpha - \beta & \alpha - \beta \\ 2\beta - 2\alpha & 2\beta - \alpha \end{bmatrix}$$

21. T

22. T

23. F. The matrix is similar to a diagonal matrix but it is unique up to permutation of the diagonal entries.

24. F. The matrix can still have n linearly independent eigenvectors. An example is the $n \times n$ identity matrix.

25. T

26. T

27. T

28. T

29. T

30. F. The matrix

$$\begin{bmatrix} 1 & 1 \\ 0 & 1 \end{bmatrix}$$

is invertible but not diagonalizable.

31. T

32. T

33. T

34. F. The zero vector has to be added.

35. T

36. T

37. T

38. T

39. T

40. T

6 Inner Product Spaces

Exercise Set 6.1

In Section 6.1, the geometry in the plane and three space is extended to the Euclidean spaces $\mathbb{R}^n$. The dot product of two vectors is of central importance. The dot product of $\mathbf{u}$ and $\mathbf{v}$ in $\mathbb{R}^n$ is the sum of the products of corresponding components. So

$$\mathbf{u} \cdot \mathbf{v} = \begin{bmatrix} u_1 \\ u_2 \\ \vdots \\ u_n \end{bmatrix} \cdot \begin{bmatrix} v_1 \\ v_2 \\ \vdots \\ v_n \end{bmatrix} = u_1 v_1 + u_2 v_2 + \cdots + u_n v_n.$$

In $\mathbb{R}^3$, then

$$\mathbf{u} \cdot \mathbf{u} = u_1^2 + u_2^2 + u_3^2, \quad \text{so that} \quad \sqrt{\mathbf{u} \cdot \mathbf{u}} = \sqrt{u_1^2 + u_2^2 + u_3^2},$$

which is the length of the vector $\mathbf{u}$, called the norm of $\mathbf{u}$ and denoted by $\|\mathbf{u}\|$. For example, the length of a vector in $\mathbb{R}^n$ is then defined as $\|\mathbf{u}\| = \sqrt{u_1^2 + u_2^2 + \cdots + u_n^2}$ and the distance between two vectors is $\|\mathbf{u} - \mathbf{v}\| = \sqrt{\mathbf{u} \cdot \mathbf{u}}$. Other properties and definitions presented in the section are:

- $\|c\mathbf{u}\| = |c| \|\mathbf{u}\|$

- $\mathbf{u} \cdot \mathbf{u} \geq 0$ and $\mathbf{u} \cdot \mathbf{u} = 0$ if and only if $\mathbf{u} = \mathbf{0}$

- $\mathbf{u} \cdot \mathbf{v} = \mathbf{v} \cdot \mathbf{u}$

- $\mathbf{u} \cdot (\mathbf{v} + \mathbf{w}) = \mathbf{u} \cdot \mathbf{v} + \mathbf{u} \cdot \mathbf{w}$

- $(c\mathbf{u}) \cdot \mathbf{v} = c(\mathbf{u} \cdot \mathbf{v})$

- The cosine of the angle between two vectors is given by $\cos \theta = \dfrac{\mathbf{u} \cdot \mathbf{v}}{\|\mathbf{u}\| \|\mathbf{v}\|}$.

- Two vectors are perpendicular if and only if $\mathbf{u} \cdot \mathbf{v} = 0$.

- A unit vector in the direction of $\mathbf{u}$ is $\dfrac{1}{\|\mathbf{u}\|} \mathbf{u}$. For example, if $\mathbf{u} = \begin{bmatrix} 1 \\ 2 \\ -3 \end{bmatrix}$, then $\|\mathbf{u}\| = \sqrt{1 + 4 + 9} = \sqrt{14}$,

  so a vector of length 1 and in the direction of $\mathbf{u}$ is $\dfrac{1}{\sqrt{14}} \begin{bmatrix} 1 \\ 2 \\ -3 \end{bmatrix}$.

A very useful observation that is used in this exercise set and later ones is that when a set of vectors $\{\mathbf{v_1}, \mathbf{v_2}, \ldots, \mathbf{v_n}\}$ is pairwise orthogonal, which means $\mathbf{v_i} \cdot \mathbf{v_j} = 0$ whenever $i \neq j$, then using the properties of dot product

$$\mathbf{v_i} \cdot (c_1 \mathbf{v_1} + c_2 \mathbf{v_2} + \cdots + c_i \mathbf{v_i} + \cdots + c_n \mathbf{v_n}) = c_1 (\mathbf{v_i} \cdot \mathbf{v_1}) + c_2 (\mathbf{v_i} \cdot \mathbf{v_2}) + c_i (\mathbf{v_i} \cdot \mathbf{v_i}) + \cdots + c_n (\mathbf{v_i} \cdot \mathbf{v_n})$$
$$= 0 + \cdots + 0 + c_i (\mathbf{v_i} \cdot \mathbf{v_i}) + 0 + \cdots + 0$$
$$= c_i (\mathbf{v_i} \cdot \mathbf{v_i}) = c_i \|\mathbf{v_i}\|^2.$$

■ Solutions to Odd Exercises

1. $\mathbf{u} \cdot \mathbf{v} = (0)(1) + (1)(-1) + (3)(2) = 5$

3.
$$\mathbf{u} \cdot (\mathbf{v} + 2\mathbf{w}) = \mathbf{u} \cdot \begin{bmatrix} 3 \\ 1 \\ -4 \end{bmatrix} = 0 + 1 - 12 = -11$$

5. $\| \mathbf{u} \| = \sqrt{1^2 + 5^2} = \sqrt{26}$

7. Divide each component of the vector by the norm of the vector, so that $\frac{1}{\sqrt{26}} \begin{bmatrix} 1 \\ 5 \end{bmatrix}$ is a unit vector in the direction of $\mathbf{u}$.

9. $\frac{10}{\sqrt{5}} \begin{bmatrix} 2 \\ 1 \end{bmatrix}$

11. $\| \mathbf{u} \| = \sqrt{(-3)^2 + (-2)^2 + 3^2} = \sqrt{22}$

13. $\frac{1}{\sqrt{22}} \begin{bmatrix} -3 \\ -2 \\ 3 \end{bmatrix}$

15. $-\frac{3}{\sqrt{11}} \begin{bmatrix} -1 \\ -1 \\ -3 \end{bmatrix} = \frac{3}{\sqrt{11}} \begin{bmatrix} 1 \\ 1 \\ 3 \end{bmatrix}$

17. Since two vectors in $\mathbb{R}^2$ are orthogonal if and only if their dot product is zero, solving $\begin{bmatrix} c \\ 3 \end{bmatrix} \cdot \begin{bmatrix} -1 \\ 2 \end{bmatrix} = 0$, gives $-c + 6 = 0$, that is, $c = 6$.

19. The pairs of vectors with dot product equaling 0 are $\mathbf{v}_1 \perp \mathbf{v}_2$, $\mathbf{v}_1 \perp \mathbf{v}_4$, $\mathbf{v}_1 \perp \mathbf{v}_5$, $\mathbf{v}_2 \perp \mathbf{v}_3$, $\mathbf{v}_3 \perp \mathbf{v}_4$, and $\mathbf{v}_3 \perp \mathbf{v}_5$.

21. Since $\mathbf{v}_3 = -\mathbf{v}_1$, the vectors $\mathbf{v}_1$ and $\mathbf{v}_3$ are in opposite directions.

23. $\mathbf{w} = \begin{bmatrix} 2 \\ 0 \end{bmatrix}$

25. $\mathbf{w} = \frac{3}{2} \begin{bmatrix} 3 \\ 1 \end{bmatrix}$

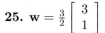

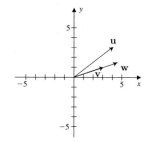

27. $\mathbf{w} = \frac{1}{6} \begin{bmatrix} 5 \\ 2 \\ 1 \end{bmatrix}$

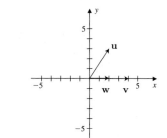

29. Let $\mathbf{u}$ be a vector in $\mathbf{span}\{\mathbf{u}_1, \mathbf{u}_2, \cdots, \mathbf{u}_n\}$. Then there exist scalars $c_1, c_2, \cdots, c_n$ such that

$$\mathbf{u} = c_1 \mathbf{u}_1 + c_2 \mathbf{u}_2 + \cdots + c_n \mathbf{u}_n.$$

Using the distributive property of the dot product gives

$$\begin{aligned}
\mathbf{v} \cdot \mathbf{u} &= \mathbf{v} \cdot (c_1 \mathbf{u}_1 + c_2 \mathbf{u}_2 + \ldots + c_n \mathbf{u}_n) \\
&= c_1 (\mathbf{v} \cdot \mathbf{u}_1) + c_2 (\mathbf{v} \cdot \mathbf{u}_2) + \cdots + c_n (\mathbf{v} \cdot \mathbf{u}_n) \\
&= c_1 (0) + c_2 (0) + \cdots + c_n (0) = 0.
\end{aligned}$$

31. Consider the equation

$$c_1\mathbf{v_1} + c_2\mathbf{v_2} + \cdots + c_n\mathbf{v_n} = \mathbf{0}.$$

We need to show that the only solution to this equation is the trivial solution $c_1 = c_2 = \cdots = c_n = 0$. Since

$$\mathbf{v_1} \cdot (c_1\mathbf{v_1} + c_2\mathbf{v_2} + \cdots + c_n\mathbf{v_n}) = \mathbf{v_1} \cdot \mathbf{0}, \text{ we have that } c_1(\mathbf{v_1} \cdot \mathbf{v_1}) + c_2(\mathbf{v_1} \cdot \mathbf{v_2}) + \cdots + c_n(\mathbf{v_1} \cdot \mathbf{v_n}) = 0.$$

Since S is an orthogonal set of vectors, $\mathbf{v_i} \cdot \mathbf{v_j} = 0$, whenever $i \neq j$, so this last equation reduces to $c_1||\mathbf{v_1}||^2 = 0$. Now since the vectors are nonzero their lengths are positive, so $||\mathbf{v_1}|| \neq 0$ and hence, $c_1 = 0$. In a similar way we have that $c_2 = c_3 = \cdots = c_n = 0$. Hence, S is linearly independent.

33. For any vector $\mathbf{w}$, the square of the norm and the dot product are related by the equation $||\mathbf{w}||^2 = \mathbf{w} \cdot \mathbf{w}$. Then applying this to the vectors $\mathbf{u} + \mathbf{v}$ and $\mathbf{u} - \mathbf{v}$ gives

$$
\begin{aligned}
||\mathbf{u} + \mathbf{v}||^2 + ||\mathbf{u} - \mathbf{v}||^2 &= (\mathbf{u} + \mathbf{v}) \cdot (\mathbf{u} + \mathbf{v}) + (\mathbf{u} - \mathbf{v}) \cdot (\mathbf{u} - \mathbf{v}) \\
&= \mathbf{u} \cdot \mathbf{u} + 2\mathbf{u} \cdot \mathbf{v} + \mathbf{v} \cdot \mathbf{v} + \mathbf{u} \cdot \mathbf{u} - 2\mathbf{u} \cdot \mathbf{v} + \mathbf{v} \cdot \mathbf{v} \\
&= 2||\mathbf{u}||^2 + 2||\mathbf{v}||^2.
\end{aligned}
$$

35. If the column vectors of A form an orthogonal set, then the row vectors of A^t are orthogonal to the column vectors of A. Consequently, $(A^t A)_{ij} = 0$ if $i \neq j$. If $i = j$, then $(A^t A)_{ii} = ||\mathbf{A_i}||^2$. Thus,

$$
A^t A = \begin{bmatrix}
||\mathbf{A_1}||^2 & 0 & \cdots & 0 \\
0 & ||\mathbf{A_2}||^2 & 0 & \vdots \\
\vdots & 0 & \ddots & 0 \\
0 & \cdots & 0 & ||\mathbf{A_n}||^2
\end{bmatrix}.
$$

37. Suppose that $(A\mathbf{u}) \cdot \mathbf{v} = \mathbf{u} \cdot (A\mathbf{v})$ for all $\mathbf{u}$ and $\mathbf{v}$ in $\mathbb{R}^n$. By Exercise 36, $\mathbf{u} \cdot (A\mathbf{v}) = (A^t\mathbf{u}) \cdot \mathbf{v}$. Thus,

$$(A^t\mathbf{u}) \cdot \mathbf{v} = (A\mathbf{u}) \cdot \mathbf{v}$$

for all $\mathbf{u}$ and $\mathbf{v}$ in $\mathbb{R}^n$. Let $\mathbf{u} = \mathbf{e_i}$ and $\mathbf{v} = \mathbf{e_j}$, so $(A^t)_{ij} = A_{ij}$. Hence $A^t = A$, so A is symmetric. For the converse, suppose that $A = A^t$. Then by Exercise 36,

$$\mathbf{u} \cdot (A\mathbf{v}) = (A^t\mathbf{u}) \cdot \mathbf{v} = (A\mathbf{u}) \cdot \mathbf{v}.$$

Exercise Set 6.2

Inner products are generalizations of the dot product on the Euclidean spaces. An inner product on the vector space V is a mapping from $V \times V$, that is the input is a pair of vectors, to $\mathbb{R}$, so the output is a number. An inner product must satisfy all the same properties of the dot product discussed in Section 6.1. So to determine if a mapping on $V \times V$ defines an inner product we must verify that:

- $\langle \mathbf{u}, \mathbf{u} \rangle \geq 0$ and $\langle \mathbf{u}, \mathbf{u} \rangle = 0$ if and only if $\mathbf{u} = \mathbf{0}$

- $\langle \mathbf{u}, \mathbf{v} \rangle = \langle \mathbf{v}, \mathbf{u} \rangle$

- $\langle \mathbf{u} + \mathbf{v}, \mathbf{w} \rangle = \langle \mathbf{u}, \mathbf{w} \rangle + \langle \mathbf{v}, \mathbf{w} \rangle$

- $\langle c\mathbf{u}, \mathbf{v} \rangle = c \langle \mathbf{u}, \mathbf{v} \rangle$

In addition the definition of length, distance and angle between vectors are given in the same way with dot product replaced with inner product. So

$$||\mathbf{v}|| = \sqrt{\langle \mathbf{v}, \mathbf{v} \rangle} \quad \text{and} \quad \cos\theta = \frac{\langle \mathbf{u}, \mathbf{v} \rangle}{||\mathbf{u}||||\mathbf{v}||}.$$

Also two vectors in an inner product space are orthogonal if and only if $\langle \mathbf{u}, \mathbf{v} \rangle = 0$ and a set $S = \{\mathbf{v_1}, \mathbf{v_2}, \ldots, \mathbf{v_k}\}$ is orthogonal if the vectors are pairwise orthogonal. That is, $\langle \mathbf{v_i}, \mathbf{v_j} \rangle = 0$, whenever $i \neq j$. If in addition, for each $i = 1, \ldots, k$, $\|\mathbf{v_i}\| = 1$, then S is orthonormal. Useful in the exercise set is the fact that every orthogonal set of nonzero vectors in an inner product space is linearly independent. If $B = \{\mathbf{v_1}, \mathbf{v_2}, \ldots, \mathbf{v_n}\}$ is an orthogonal basis for an inner product space, then for every vector $\mathbf{v}$ the scalars needed to write the vector in terms of the basis vectors are given explicitly by the inner product, so that

$$\mathbf{v} = \frac{\langle \mathbf{v}, \mathbf{v_1} \rangle}{\langle \mathbf{v_1}, \mathbf{v_1} \rangle} \mathbf{v_1} + \frac{\langle \mathbf{v}, \mathbf{v_2} \rangle}{\langle \mathbf{v_2}, \mathbf{v_2} \rangle} \mathbf{v_2} + \cdots + \frac{\langle \mathbf{v}, \mathbf{v_n} \rangle}{\langle \mathbf{v_n}, \mathbf{v_n} \rangle} \mathbf{v_n}.$$

If B is orthonormal, then the expansion is

$$\mathbf{v} = \langle \mathbf{v}, \mathbf{v_1} \rangle \mathbf{v_1} + \langle \mathbf{v}, \mathbf{v_2} \rangle \mathbf{v_2} + \cdots + \langle \mathbf{v}, \mathbf{v_n} \rangle \mathbf{v_n}.$$

■ Solutions to Odd Exercises

1. Let $\mathbf{u} = \begin{bmatrix} u_1 \\ u_2 \end{bmatrix}$. To examine the definition given for $\langle \mathbf{u}, \mathbf{v} \rangle$, we will first let $\mathbf{u} = \mathbf{v}$. Then

$$\langle \mathbf{u}, \mathbf{u} \rangle = u_1^2 - 2u_1 u_2 - 2u_2 u_1 + 3u_2^2 = (u_1 - 3u_2)(u_1 + u_2).$$

Since $\langle \mathbf{u}, \mathbf{u} \rangle = 0$ if and only if $u_1 = 3u_2$ or $u_1 = u_2$, then V is not an inner product space. For example, if $u_1 = 3$ and $u_2 = 1$, then $\langle \mathbf{u}, \mathbf{u} \rangle = 0$ but $\mathbf{u}$ is not the zero vector.

3. In this case $\langle \mathbf{u}, \mathbf{u} \rangle = 0$ if and only if $\mathbf{u}$ is the zero vector and $\langle \mathbf{u}, \mathbf{v} \rangle = \langle \mathbf{v}, \mathbf{u} \rangle$ for all pairs of vectors. To check whether the third requirement holds, we have that

$$\langle \mathbf{u} + \mathbf{v}, \mathbf{w} \rangle = (u_1 + v_1)^2 w_1^2 + (u_2 + v_2)^2 w_2^2$$

and

$$\langle \mathbf{u}, \mathbf{w} \rangle + \langle \mathbf{v}, \mathbf{w} \rangle = u_1^2 w_1^2 + u_2^2 w_2^2 + v_1^2 w_1^2 + v_2^2 w_2^2.$$

For example if $\mathbf{u} = \begin{bmatrix} 1 \\ 1 \end{bmatrix}$ and $\mathbf{v} = \begin{bmatrix} 2 \\ 2 \end{bmatrix}$, then $\langle \mathbf{u} + \mathbf{v}, \mathbf{w} \rangle = 9w_1^2 + 9w_2^2$ and $\langle \mathbf{u}, \mathbf{v} \rangle + \langle \mathbf{v}, \mathbf{w} \rangle = 5w_1^2 + 5w_2^2$. Now let $\mathbf{w} = \begin{bmatrix} -1 \\ -1 \end{bmatrix}$, so that $\langle \mathbf{u} + \mathbf{v}, \mathbf{w} \rangle = 18 \neq 10 = \langle \mathbf{u}, \mathbf{v} \rangle + \langle \mathbf{v}, \mathbf{w} \rangle$, so V is not an inner product space.

5. The four requirements for an inner product are satisfied by the dot product, so V is an inner product space.

7. The vector space V is an inner product space with the given definition for $\langle A, B \rangle$. For the first requirement $\langle A, A \rangle = \sum_{i=1}^{m} \sum_{j=1}^{n} a_{ij}^2$, which is nonnegative and 0 if and only if A is the zero matrix. Since real numbers commute, $\langle A, B \rangle = \langle B, A \rangle$. Finally,

$$\langle A + B, C \rangle = \sum_{i=1}^{m} \sum_{j=1}^{n} (a_{ij} c_{ij} + b_{ij} c_{ij}),$$

then $\langle A + B, C \rangle = \langle A, C \rangle + \langle B, C \rangle$.

9. Using the properties of the integral and the fact that the exponential function is always nonnegative, V is an inner product space.

11. The set is orthogonal provided the functions are pairwise orthogonal. That is, provided $\langle 1, \sin x \rangle = 0, \langle 1, \cos x \rangle = 0$, and $\langle \cos x, \sin x \rangle = 0$. We have

$$\langle 1, \sin x \rangle = \int_{-\pi}^{\pi} \sin x \, dx = -\cos x \, |_{-\pi}^{\pi} = 0.$$

Similarly, $\langle 1, \cos x \rangle = 0$. To show $\langle \cos x, \sin x \rangle = 0$ requires the technique of substitution, or notice that cosine is an odd function and sine is an even function, so the product is an odd function and since the integral is over a symmetric interval, the integral is 0.

13. The inner products are

$$\langle 1, 2x - 1 \rangle = \int_0^1 (2x-1)dx = 0, \left\langle 1, -x^2 + x - \frac{1}{6} \right\rangle = \int_0^1 \left(-x^2 + x - \frac{1}{6} \right) dx = 0,$$

and

$$\left\langle 2x - 1, -x^2 + x - \frac{1}{6} \right\rangle = \int_0^1 \left(-2x^3 + 3x^2 - \frac{4}{3}x + \frac{1}{6} \right) dx = 0.$$

15. a. The distance between the two functions is

$$\|f - g\| = \| -3 + 3x - x^2 \| = \sqrt{\int_0^1 (-3 + 3x - x^2)^2 dx} = \sqrt{\frac{370}{10}}.$$

b. The cosine of the angle between the functions is given by $\cos \theta = \frac{\langle f, g \rangle}{\|f\|\|g\|} = -\frac{5}{168}\sqrt{105}$.

17. a. $\| x - e^x \| = \sqrt{\frac{1}{2}e^2 - \frac{13}{6}}$ **b.** $\cos \theta = \frac{2\sqrt{3}}{\sqrt{2e^2 - 2}}$ **19. a.** $\| 2x^2 - 4 \| = 2\sqrt{5}$ **b.** $\cos \theta = -\frac{2}{3}$

21. a. To find the distance between the matrices, we have that

$$\| A - B \| = \left\| \begin{bmatrix} 1 & 2 \\ 2 & -1 \end{bmatrix} - \begin{bmatrix} 2 & 1 \\ 1 & 3 \end{bmatrix} \right\| = \left\| \begin{bmatrix} -1 & 1 \\ 1 & -4 \end{bmatrix} \right\| = \sqrt{\left\langle \begin{bmatrix} -1 & 1 \\ 1 & -4 \end{bmatrix}, \begin{bmatrix} -1 & 1 \\ 1 & -4 \end{bmatrix} \right\rangle}$$

$$= \sqrt{\mathbf{tr}\left(\begin{bmatrix} -1 & 1 \\ 1 & -4 \end{bmatrix} \begin{bmatrix} -1 & 1 \\ 1 & -4 \end{bmatrix} \right)} = \sqrt{\mathbf{tr}\left(\begin{bmatrix} 2 & -5 \\ -5 & 17 \end{bmatrix} \right)} = \sqrt{19}.$$

b. The cosine of the angle between the matrices is given by $\cos \theta = \frac{\langle A, B \rangle}{\|A\|\|B\|} = \frac{3}{5\sqrt{6}}$.

23. a. $\| A - B \| = \sqrt{\mathbf{tr} \begin{bmatrix} 8 & 0 & 8 \\ 0 & 3 & 4 \\ 8 & 4 & 14 \end{bmatrix}} = \sqrt{25} = 5$ **b.** $\cos \theta = \frac{26}{\sqrt{38}\sqrt{39}}$

25. Since $\begin{bmatrix} x \\ y \end{bmatrix}$ is orthogonal to $\begin{bmatrix} 2 \\ 3 \end{bmatrix}$ if and only if $\begin{bmatrix} x \\ y \end{bmatrix} \cdot \begin{bmatrix} 2 \\ 3 \end{bmatrix} = 0$, which is true if and only if $2x + 3y = 0$,

so the set of vectors that are orthogonal to $\begin{bmatrix} 2 \\ 3 \end{bmatrix}$ is $\left\{ \begin{bmatrix} x \\ y \end{bmatrix} \middle| 2x + 3y = 0 \right\}$. Notice that the set describes a line.

27. Since $\begin{bmatrix} x \\ y \\ x \end{bmatrix} \cdot \begin{bmatrix} 2 \\ -3 \\ 1 \end{bmatrix} = 0$ if and only if $2x - 3y + z = 0$, the set of all vectors orthogonal to $\mathbf{n} = \begin{bmatrix} 2 \\ -3 \\ 1 \end{bmatrix}$

is $\left\{ \begin{bmatrix} x \\ y \\ z \end{bmatrix} \middle| 2x - 3y + z = 0 \right\}$. This set describes the plane with normal vector $\mathbf{n}$.

29. a. $\langle x^2, x^3 \rangle = \int_0^1 x^5 dx = \frac{1}{6}$ **b.** $\langle e^x, e^{-x} \rangle = \int_0^1 dx = 1$ **c.** $\| 1 \| = \sqrt{\int_0^1 dx} = 1$,

$\| x \| = \sqrt{\int_0^1 x^2 dx} = \frac{\sqrt{3}}{3}$ **d.** $\cos\theta = \frac{3}{2\sqrt{3}}$ **e.** $\| 1 - x \| = \frac{\sqrt{3}}{3}$

31. If f is an even function and g is an odd function, then fg is an odd function. Since the inner product is defined as the integral over a symmetric interval $[-a, a]$, then

$$\langle f, g \rangle = \int_{-a}^{a} f(x)g(x)dx = 0,$$

so f and g are orthogonal.

33. Using the scalar property of inner products, we have that

$$\langle c_1 \mathbf{u_1}, c_2 \mathbf{u_2} \rangle = c_1 \langle \mathbf{u_1}, c_2 \mathbf{u_2} \rangle = c_1 c_2 \langle \mathbf{u_1}, \mathbf{u_2} \rangle.$$

Since $\mathbf{u_1}$ and $\mathbf{u_2}$ are orthogonal, $\langle \mathbf{u_1}, \mathbf{u_2} \rangle = 0$ and hence, $\langle c_1 \mathbf{u_1}, c_2 \mathbf{u_2} \rangle = 0$. Therefore, $c_1 \mathbf{u_1}$ and $c_2 \mathbf{u_2}$ are orthogonal.

Exercise Set 6.3

Every finite dimensional inner product space has an orthogonal basis. Given any basis B the Gram-Schmidt process is a method for constructing an orthogonal basis from B. The construction process involves projecting one vector onto another. The orthogonal projection of $\mathbf{u}$ onto $\mathbf{v}$ is

$$\text{proj}_{\mathbf{v}} \mathbf{u} = \frac{\mathbf{u} \cdot \mathbf{v}}{\mathbf{v} \cdot \mathbf{v}} \mathbf{v} = \frac{\langle \mathbf{u}, \mathbf{v} \rangle}{\langle \mathbf{v}, \mathbf{v} \rangle} \mathbf{v}.$$

Notice that the vectors $\text{proj}_{\mathbf{v}} \mathbf{u}$ and $\mathbf{u} - \text{proj}_{\mathbf{v}} \mathbf{u}$ are orthogonal.

Let $B = \{\mathbf{v_1}, \mathbf{v_2}, \mathbf{v_3}\}$ be the basis for $\mathbb{R}^3$ given by

$$\mathbf{v_1} = \begin{bmatrix} 1 \\ 2 \\ -1 \end{bmatrix}, \mathbf{v_2} = \begin{bmatrix} 1 \\ 0 \\ 1 \end{bmatrix}, \mathbf{v_3} = \begin{bmatrix} 1 \\ -1 \\ 1 \end{bmatrix}.$$

Notice that B is not an orthogonal basis, since $\mathbf{v_1}$ and $\mathbf{v_3}$ are not orthogonal, even though $\mathbf{v_1}$ and $\mathbf{v_2}$ are orthogonal. Also, $\mathbf{v_2}$ and $\mathbf{v_3}$ are not orthogonal.

Gram-Schmidt Process to Covert B to the Orthogonal Basis $\{\mathbf{w_1}, \mathbf{w_2}, \mathbf{w_3}\}$.

- $\mathbf{w_1} = \mathbf{v_1} = \begin{bmatrix} 1 \\ 2 \\ -1 \end{bmatrix}$

- $\mathbf{w_2} = \mathbf{v_2} - \text{proj}_{\mathbf{w_1}} \mathbf{v_2} = \mathbf{v_2} - \frac{\langle \mathbf{v_2}, \mathbf{w_1} \rangle}{\langle \mathbf{w_1}, \mathbf{w_1} \rangle} \mathbf{w_1} = \begin{bmatrix} 1 \\ 0 \\ 1 \end{bmatrix} - \frac{\begin{bmatrix} 1 \\ 0 \\ 1 \end{bmatrix} \cdot \begin{bmatrix} 1 \\ 2 \\ -1 \end{bmatrix}}{\begin{bmatrix} 1 \\ 2 \\ -1 \end{bmatrix} \cdot \begin{bmatrix} 1 \\ 2 \\ -1 \end{bmatrix}} \begin{bmatrix} 1 \\ 2 \\ -1 \end{bmatrix} = \begin{bmatrix} 1 \\ 0 \\ 1 \end{bmatrix}$

- $\mathbf{w_3} = \mathbf{v_3} - \text{proj}_{\mathbf{w_1}} \mathbf{v_3} - \text{proj}_{\mathbf{w_2}} \mathbf{v_3} = \mathbf{v_3} - \frac{\langle \mathbf{v_3}, \mathbf{w_1} \rangle}{\langle \mathbf{w_1}, \mathbf{w_1} \rangle} \mathbf{w_1} - \frac{\langle \mathbf{v_3}, \mathbf{w_2} \rangle}{\langle \mathbf{w_2}, \mathbf{w_2} \rangle} \mathbf{w_2}$

$$= \begin{bmatrix} 1 \\ -1 \\ 1 \end{bmatrix} - \frac{\begin{bmatrix} 1 \\ -1 \\ 1 \end{bmatrix} \cdot \begin{bmatrix} 1 \\ 2 \\ -1 \end{bmatrix}}{\begin{bmatrix} 1 \\ 2 \\ -1 \end{bmatrix} \cdot \begin{bmatrix} 1 \\ 2 \\ -1 \end{bmatrix}} \begin{bmatrix} 1 \\ 2 \\ -1 \end{bmatrix} - \frac{\begin{bmatrix} 1 \\ -1 \\ 1 \end{bmatrix} \cdot \begin{bmatrix} 1 \\ 0 \\ 1 \end{bmatrix}}{\begin{bmatrix} 1 \\ 0 \\ 1 \end{bmatrix} \cdot \begin{bmatrix} 1 \\ 0 \\ 1 \end{bmatrix}} \begin{bmatrix} 1 \\ 0 \\ 1 \end{bmatrix} = \begin{bmatrix} 1/3 \\ -1/3 \\ -1/3 \end{bmatrix}$$

An orthonormal basis is found by dividing each of the vectors in an orthogonal basis by their norms.

■ Solutions to Odd Exercises

1. a. $\text{proj}_{\mathbf{v}}\mathbf{u} = \begin{bmatrix} -3/2 \\ 3/2 \end{bmatrix}$ **b.** Since $\mathbf{u} - \text{proj}_{\mathbf{v}}\mathbf{u} = \begin{bmatrix} 1/2 \\ 1/2 \end{bmatrix}$, we have that

$$\mathbf{v} \cdot (\mathbf{u} - \text{proj}_{\mathbf{v}}\mathbf{u}) = \begin{bmatrix} -1 \\ 1 \end{bmatrix} \cdot \begin{bmatrix} 1/2 \\ 1/2 \end{bmatrix} = 0,$$

so the dot product is 0 and hence, the two vectors are orthogonal.

3. a. $\text{proj}_{\mathbf{v}}\mathbf{u} = \begin{bmatrix} -3/5 \\ -6/5 \end{bmatrix}$
b. Since

$$\mathbf{u} - \text{proj}_{\mathbf{v}}\mathbf{u} = \begin{bmatrix} 8/5 \\ -4/5 \end{bmatrix},$$

we have that

$$\mathbf{v} \cdot (\mathbf{u} - \text{proj}_{\mathbf{v}}\mathbf{u}) = \begin{bmatrix} 1 \\ 2 \end{bmatrix} \cdot \begin{bmatrix} 8/5 \\ -4/5 \end{bmatrix} = 0,$$

so the dot product is 0 and hence, the two vectors are orthogonal.

5. a. $\text{proj}_{\mathbf{v}}\mathbf{u} = \begin{bmatrix} -4/3 \\ 4/3 \\ 4/3 \end{bmatrix}$ **b.** Since

$$\mathbf{u} - \text{proj}_{\mathbf{v}}\mathbf{u} = \begin{bmatrix} 1/3 \\ 5/3 \\ -4/3 \end{bmatrix},$$

we have that

$$\mathbf{v} \cdot (\mathbf{u} - \text{proj}_{\mathbf{v}}\mathbf{u}) = \begin{bmatrix} 1 \\ -1 \\ -1 \end{bmatrix} \cdot \begin{bmatrix} 1/3 \\ 5/3 \\ -4/3 \end{bmatrix} = 0$$

so the dot product is 0 and hence, the two vectors are orthogonal.

7. a. $\text{proj}_{\mathbf{v}}\mathbf{u} = \begin{bmatrix} 0 \\ 0 \\ -1 \end{bmatrix}$ **b.** Since

$$\mathbf{u} - \text{proj}_{\mathbf{v}}\mathbf{u} = \begin{bmatrix} 1 \\ -1 \\ 0 \end{bmatrix},$$

we have that

$$\mathbf{v} \cdot (\mathbf{u} - \text{proj}_{\mathbf{v}}\mathbf{u}) = \begin{bmatrix} 0 \\ 0 \\ 1 \end{bmatrix} \cdot \begin{bmatrix} 1 \\ -1 \\ 0 \end{bmatrix} = 0,$$

so the dot product is 0 and hence, the two vectors are orthogonal.

9. a. $\text{proj}_{\mathbf{q}}\mathbf{p} = \frac{5}{4}x - \frac{5}{12}$ **b.** Since

$$\mathbf{p} - \text{proj}_{\mathbf{q}}\mathbf{p} = x^2 - \frac{9}{4}x + \frac{17}{12},$$

we have that

$$\langle \mathbf{q}, \mathbf{p} - \text{proj}_{\mathbf{q}}\mathbf{p} \rangle$$
$$= \int_0^1 (3x - 1)\left(x^2 - \frac{9}{4}x + \frac{17}{12}\right) dx = 0$$

so the dot product is 0 and hence, the two vectors are orthogonal.

11. a. $\text{proj}_{\mathbf{q}}\mathbf{p} = \frac{-7}{4}x^2 + \frac{7}{4}$ **b.** Since

$$\mathbf{p} - \text{proj}_{\mathbf{q}}\mathbf{p} = \frac{15}{4}x^2 - \frac{3}{4},$$

we have that

$$\langle \mathbf{q}, \mathbf{p} - \text{proj}_{\mathbf{q}}\mathbf{p} \rangle$$
$$= \int_0^1 (x^2 - 1)\left(\frac{15}{4}x^2 - \frac{3}{4}\right) dx = 0$$

so the dot product is 0, and hence, the two vectors are orthogonal.

13. Let $B = \{\mathbf{v_1}, \mathbf{v_2}\}$ and denote the orthogonal basis by $\{\mathbf{w_1}, \mathbf{w_2}\}$. Then

$$\mathbf{w_1} = \mathbf{v_1} = \begin{bmatrix} 1 \\ -1 \end{bmatrix}, \mathbf{w_2} = \mathbf{v_2} - \text{proj}_{\mathbf{w_1}}\mathbf{v_2} = \begin{bmatrix} -1 \\ -1 \end{bmatrix}.$$

To obtain an orthonormal basis, divide each vector in the orthogonal basis by their norm. Since each has length $\sqrt{2}$, an orthonormal basis is $\left\{ \frac{1}{\sqrt{2}}\begin{bmatrix} 1 \\ -1 \end{bmatrix}, \frac{1}{\sqrt{2}}\begin{bmatrix} -1 \\ -1 \end{bmatrix} \right\}$.

15. Let $B = \{\mathbf{v_1}, \mathbf{v_2}, \mathbf{v_3}\}$ and denote the orthogonal basis by $\{\mathbf{w_1}, \mathbf{w_2}, \mathbf{w_3}\}$. Then

$$\mathbf{w_1} = \mathbf{v_1} = \begin{bmatrix} 1 \\ 0 \\ 1 \end{bmatrix}, \mathbf{w_2} = \mathbf{v_2} - \text{proj}_{\mathbf{w_1}} \mathbf{v_2} = \begin{bmatrix} 1 \\ 2 \\ -1 \end{bmatrix}, \mathbf{w_3} = \mathbf{v_3} - \text{proj}_{\mathbf{w_1}} \mathbf{v_3} - \text{proj}_{\mathbf{w_2}} \mathbf{v_3} = \begin{bmatrix} 1 \\ -1 \\ -1 \end{bmatrix}.$$

To obtain an orthonormal basis, divide each vector in the orthogonal basis by their norm. This gives the orthonormal basis $\left\{ \frac{1}{\sqrt{2}} \begin{bmatrix} 1 \\ 0 \\ 1 \end{bmatrix}, \frac{1}{\sqrt{6}} \begin{bmatrix} 1 \\ 2 \\ -1 \end{bmatrix}, \frac{1}{\sqrt{3}} \begin{bmatrix} 1 \\ -1 \\ -1 \end{bmatrix} \right\}$.

17. $\left\{ \sqrt{3}(x-1),\ 3x-1,\ 6\sqrt{5}(x^2 - x + \frac{1}{6}) \right\}$

19. An orthonormal basis for $\mathbf{span}(W)$ is given by

$$\left\{ \frac{1}{\sqrt{3}} \begin{bmatrix} 1 \\ 1 \\ 1 \end{bmatrix}, \frac{1}{\sqrt{6}} \begin{bmatrix} 2 \\ -1 \\ -1 \end{bmatrix} \right\}.$$

21. An orthonormal basis for $\mathbf{span}(W)$ is given by

$$\left\{ \frac{1}{\sqrt{6}} \begin{bmatrix} -1 \\ -2 \\ 0 \\ 1 \end{bmatrix}, \frac{1}{\sqrt{6}} \begin{bmatrix} -2 \\ 1 \\ -1 \\ 0 \end{bmatrix}, \frac{1}{\sqrt{6}} \begin{bmatrix} 1 \\ 0 \\ -2 \\ 1 \end{bmatrix} \right\}.$$

23. An orthonormal basis for $\mathbf{span}(W)$ is given by

$$\left\{ \sqrt{3}x,\ -3x + 2 \right\}.$$

25. An orthonormal basis for $\mathbf{span}(W)$ is given by $\left\{ \frac{1}{\sqrt{3}} \begin{bmatrix} 1 \\ 0 \\ 1 \\ 1 \end{bmatrix}, \frac{1}{\sqrt{3}} \begin{bmatrix} 0 \\ 1 \\ -1 \\ 1 \end{bmatrix} \right\}.$

27. Let $\mathbf{v}$ be a vector in V and $B = \{\mathbf{u_1}, \mathbf{u_2}, \dots, \mathbf{u_n}\}$ an orthonormal basis for V. Then there exist scalars $c_1, c_2, \dots, c_n$ such that $\mathbf{v} = c_1 \mathbf{u_1} + c_2 \mathbf{u_2} + \cdots + c_n \mathbf{u_n}$. Then

$$||\mathbf{v}||^2 = \mathbf{v} \cdot \mathbf{v} = c_1^2 (\mathbf{u_1} \cdot \mathbf{u_1}) + c_2^2 (\mathbf{u_2} \cdot \mathbf{u_2}) + \cdots + c_n^2 (\mathbf{u_n} \cdot \mathbf{u_n}).$$

Since B is orthonormal each vector in B has norm one, $1 = ||\mathbf{u_i}||^2 = \mathbf{u_i} \cdot \mathbf{u_i}$ and they are pairwise orthogonal, so $\mathbf{u_1} \cdot \mathbf{u_j} = 0$, for $i \neq j$. Hence,

$$||\mathbf{v}||^2 = c_1^2 + c_2^2 + \cdots + c_n^2 = |\mathbf{v} \cdot \mathbf{u_1}|^2 + \cdots + |\mathbf{v} \cdot \mathbf{u_n}|^2.$$

29. Let

$$A = \begin{bmatrix} a_{11} & a_{12} & \cdots & a_{1n} \\ a_{21} & a_{22} & \cdots & a_{2n} \\ \vdots & \vdots & \vdots & \vdots \\ a_{n1} & a_{n2} & \cdots & a_{nn} \end{bmatrix} \quad \text{and} \quad A^t = \begin{bmatrix} a_{11} & a_{21} & \cdots & a_{n1} \\ a_{12} & a_{22} & \cdots & a_{n2} \\ \vdots & \vdots & \vdots & \vdots \\ a_{1n} & a_{2n} & \cdots & a_{nn} \end{bmatrix}.$$

Suppose that the columns of A form an orthonormal set. So that $\sum_{k=1}^{n} a_{ki} a_{kj} = \begin{cases} 0 \text{ if } i \neq j \\ 1 \text{ if } i = j \end{cases}$. Observe that

the quantity on the left hand side of this equation is the i, j entry of $A^t A$, so that $(A^t A)_{ij} = \begin{cases} 0 \text{ if } i \neq j \\ 1 \text{ if } i = j \end{cases}$

and hence, $A^t A = I$. Conversely, if $A^t A = I$, then A has orthonormal columns.

31. Recall that $||A\mathbf{x}|| = \sqrt{A\mathbf{x} \cdot A\mathbf{x}}$. Since the column vectors of A are orthonormal, by Exercise 29, we have that $A^t A = I$. By Exercise 30, we have that

$$A\mathbf{x} \cdot A\mathbf{x} = (A^t A\mathbf{x}) \cdot \mathbf{x} = \mathbf{x}^t \cdot (A^t A\mathbf{x}) = \mathbf{x} \cdot \mathbf{x}.$$

Since the left hand side is $||A\mathbf{x}||^2$, we have that $||A\mathbf{x}||^2 = \mathbf{x} \cdot \mathbf{x} = ||\mathbf{x}||^2$ and hence, $||A\mathbf{x}|| = ||\mathbf{x}||$.

33. By Exercise 32, $A\mathbf{x} \cdot A\mathbf{y} = \mathbf{x} \cdot \mathbf{y}$. Then $A\mathbf{x} \cdot A\mathbf{y} = 0$ if and only if $\mathbf{x} \cdot \mathbf{y} = 0$.

35. Let

$$W = \{\mathbf{v} \mid \mathbf{v} \cdot \mathbf{u_i} = 0, \text{ for all } i = 1, 2, \ldots m\}.$$

Let c be a real number and $\mathbf{x}$ and $\mathbf{y}$ vectors in W, so that $\mathbf{x} \cdot \mathbf{u_i} = 0$ and $\mathbf{y} \cdot \mathbf{u_i} = 0$, for each $i = 1, \ldots, m$. Then

$$(\mathbf{x} + c\mathbf{y}) \cdot \mathbf{u_i} = \mathbf{x} \cdot \mathbf{u_i} + c(\mathbf{y} \cdot \mathbf{u_i}) = 0 + c(0) = 0,$$

for all $i = 1, 2, \ldots, n$. So $\mathbf{x} + c\mathbf{y}$ is in W and hence, W is a subspace.

37. Since for every nonzero vector $\begin{bmatrix} x \\ y \end{bmatrix}$, we have that

$$\mathbf{v}^t A\mathbf{v} = [x, y] \begin{bmatrix} 3 & 1 \\ 1 & 3 \end{bmatrix} \begin{bmatrix} x \\ y \end{bmatrix} = 3x^2 + 2xy + 3y^2 \geq (x + y)^2 \geq 0,$$

so A is positive semi-definite.

39. Since, when defined, $(BC)^t = C^t B^t$, we have that

$$\mathbf{x}^t A^t A\mathbf{x} = (A\mathbf{x})^t A\mathbf{x} = (A\mathbf{x}) \cdot (A\mathbf{x}) = ||A\mathbf{x}||^2 \geq 0,$$

so $A^t A$ is positive semi-definite.

41. Let $\mathbf{x}$ be an eigenvector of A corresponding to the eigenvalue λ, so that $A\mathbf{x} = \lambda\mathbf{x}$. Now multiply both sides by $\mathbf{x}^t$ to obtain

$$\mathbf{x}^t A\mathbf{x} = \mathbf{x}^t(\lambda\mathbf{x}) = \lambda(\mathbf{x}^t\mathbf{x}) = \lambda(\mathbf{x} \cdot \mathbf{x}) = \lambda||\mathbf{x}||^2.$$

Since A is positive definite and $\mathbf{x}$ is not the zero vector, then $\mathbf{x}^t A\mathbf{x} > 0$, so $\lambda > 0$.

Exercise Set 6.4

The orthogonality of two vectors is extended to a vector being orthogonal to a subspace of an inner product space. A vector $\mathbf{v}$ is orthogonal to a subspace W if $\mathbf{v}$ is orthogonal to every vector in W. For example, the normal vector of a plane through the origin in $\mathbb{R}^3$ is orthogonal to every vector in the plane. The orthogonal complement of a subspace is the subspace of all vectors that are orthogonal to W and is given by

$$W^\perp = \{\mathbf{v} \in V \mid \langle \mathbf{v}, \mathbf{w} \rangle = 0 \text{ for every } \mathbf{w} \in W\}.$$

If

$$W = \left\{ \begin{bmatrix} x \\ y \\ z \end{bmatrix} \middle| x - 2y + z = 0 \right\} = \left\{ y \begin{bmatrix} 1 \\ 2 \\ 0 \end{bmatrix} + z \begin{bmatrix} -1 \\ 0 \\ 1 \end{bmatrix} \middle| y, z \in \mathbb{R} \right\} = \mathbf{span} \left\{ \begin{bmatrix} 1 \\ 2 \\ 0 \end{bmatrix}, \begin{bmatrix} -1 \\ 0 \\ 1 \end{bmatrix} \right\},$$

then since the two vectors $\begin{bmatrix} 1 \\ 2 \\ 0 \end{bmatrix}$ and $\begin{bmatrix} -1 \\ 0 \\ 1 \end{bmatrix}$ are linearly independent W is a plane through the origin in

$\mathbb{R}^3$. A vector $\begin{bmatrix} a \\ b \\ c \end{bmatrix}$ is in $W^\perp$ if and only if it is orthogonal to the basis vectors, that is,

$$\begin{bmatrix} a \\ b \\ c \end{bmatrix} \cdot \begin{bmatrix} 1 \\ 2 \\ 0 \end{bmatrix} = 0 \text{ and } \begin{bmatrix} a \\ b \\ c \end{bmatrix} \cdot \begin{bmatrix} -1 \\ 0 \\ 1 \end{bmatrix} = 0 \Leftrightarrow a = c, b = -\frac{1}{2}c, c \in \mathbb{R},$$

so

$$W^\perp = \left\{ t \begin{bmatrix} 1 \\ -1/2 \\ 1 \end{bmatrix} \middle| t \in \mathbb{R} \right\}.$$

So the orthogonal complement is the line in the direction of the vector (normal vector) $\begin{bmatrix} 1 \\ -1/2 \\ 1 \end{bmatrix}$. In the previous example, we used the criteria that $\mathbf{v}$ is in the orthogonal complement of a subspace W if and only if $\mathbf{v}$ is orthogonal to each vector in a basis for $W^\perp$. Other results to remember when solving the exercises are:

- $W^\perp$ is a subspace
- $W \cap W^\perp = \{\mathbf{0}\}$
- $(W^\perp)^\perp = W$
- $\dim(V) = \dim(W) \oplus \dim(W^\perp)$

▮ Solutions to Odd Exercises

1. $W^\perp = \left\{ \begin{bmatrix} x \\ y \end{bmatrix} \middle| \begin{bmatrix} x \\ y \end{bmatrix} \cdot \begin{bmatrix} 1 \\ -2 \end{bmatrix} = 0 \right\}$

$= \left\{ \begin{bmatrix} x \\ y \end{bmatrix} \middle| x - 2y = 0 \right\}$

$= \left\{ \begin{bmatrix} x \\ y \end{bmatrix} \middle| y = \frac{1}{2}x \right\}$

$= \mathbf{span} \left\{ \begin{bmatrix} 1 \\ 1/2 \end{bmatrix} \right\}$

So the orthogonal complement is described by a line.

3.

$W^\perp = \left\{ \begin{bmatrix} x \\ y \\ z \end{bmatrix} \middle| \begin{bmatrix} x \\ y \\ z \end{bmatrix} \cdot \begin{bmatrix} 2 \\ 1 \\ -1 \end{bmatrix} = 0 \right\}$

$= \left\{ \begin{bmatrix} x \\ y \\ z \end{bmatrix} \middle| 2x + y - z = 0 \right\}$

$= \mathbf{span} \left\{ \begin{bmatrix} 1 \\ -2 \\ 0 \end{bmatrix}, \begin{bmatrix} 0 \\ 1 \\ 1 \end{bmatrix} \right\}$

So the orthogonal complement is described by a plane.

5. The orthogonal complement is the set of all vectors that are orthogonal to both $\begin{bmatrix} 2 \\ 1 \\ -1 \end{bmatrix}$ and $\begin{bmatrix} 1 \\ 2 \\ 0 \end{bmatrix}$. That is, the set of all vectors $\begin{bmatrix} x \\ y \\ z \end{bmatrix}$ satisfying

$$\begin{bmatrix} x \\ y \\ z \end{bmatrix} \cdot \begin{bmatrix} 2 \\ 1 \\ -1 \end{bmatrix} = 0 \quad \text{and} \quad \begin{bmatrix} x \\ y \\ z \end{bmatrix} \cdot \begin{bmatrix} 1 \\ 2 \\ 0 \end{bmatrix} = 0 \Leftrightarrow \begin{cases} 2x + y - z = 0 \\ x + 2y \phantom{{}- z} = 0 \end{cases}.$$

Since the solution to the linear system is $x = \frac{2}{3}z, y = -\frac{1}{3}z, z \in \mathbb{R}$, then

$$W^{\perp} = \mathbf{span} \left\{ \begin{bmatrix} 2/3 \\ -1/3 \\ 1 \end{bmatrix} \right\}.$$ Thus, the orthogonal complement is a line in three space.

7. The orthogonal complement is the set of all vectors that are orthogonal to the two given vectors. This leads to the system of equations

$$\begin{cases} 3x + y + z - w & = 0 \\ 2y + z + 2w & = 0 \end{cases}, \text{ with solution } x = -\frac{1}{6}z + \frac{2}{3}w, y = -\frac{1}{2}z - w, z, w \in \mathbb{R}.$$

Hence a vector is in $W^{\perp}$ if it only if it has the form $\begin{bmatrix} -\frac{1}{6}z + \frac{2}{3}w \\ -\frac{1}{2}z \\ z \\ w \end{bmatrix}$, for all real numbers z and w, that is,

$$W^{\perp} = \mathbf{span} \left\{ \begin{bmatrix} -1/6 \\ -1/2 \\ 1 \\ 0 \end{bmatrix}, \begin{bmatrix} 2/3 \\ -1 \\ 0 \\ 1 \end{bmatrix} \right\}.$$

9. $\left\{ \begin{bmatrix} -1/3 \\ -1/3 \\ 1 \end{bmatrix} \right\}$

11. $\left\{ \begin{bmatrix} 1/2 \\ -3/2 \\ 1 \\ 0 \end{bmatrix}, \begin{bmatrix} -1/2 \\ -1/2 \\ 0 \\ 1 \end{bmatrix} \right\}$

13. A polynomial $p(x) = ax^2 + bx + c$ is in $W^{\perp}$ if and only if $\langle p(x), x - 1 \rangle = 0$ and $\langle p(x), x^2 \rangle = 0$. Now

$$0 = \langle p(x), x - 1 \rangle = \int_0^1 (ax^3 + (b - a)x^2 + (c - b)x - c)dx \Leftrightarrow a + 2b + 6c = 0$$

and

$$0 = \langle p(x), x^2 \rangle = \int_0^1 (ax^4 + bx^3 + cx^2)dx \Leftrightarrow \frac{a}{5} + \frac{b}{4} + \frac{c}{3} = 0.$$

Since the system $\begin{cases} a + 2b + 6c & = 0 \\ \frac{a}{5} + \frac{b}{4} + \frac{c}{3} & = 0 \end{cases}$ has the solution $a = \frac{50}{9}c, b = -\frac{52}{9}, c \in \mathbb{R}$, then a basis for the orthogonal complement is $\left\{ \frac{50}{9}x^2 - \frac{52}{9}x + 1 \right\}$.

15. The set W consists of all vectors $\mathbf{w} = \begin{bmatrix} w_1 \\ w_2 \\ w_3 \\ w_4 \end{bmatrix}$ such that $w_4 = -w_1 - w_2 - w_3$, that is

$$W = \left\{ \begin{bmatrix} s \\ t \\ u \\ -s - t - u \end{bmatrix} \middle| s, t, u \in \mathbb{R} \right\} = \left\{ s \begin{bmatrix} 1 \\ 0 \\ 0 \\ -1 \end{bmatrix} + t \begin{bmatrix} 0 \\ 1 \\ 0 \\ -1 \end{bmatrix} + u \begin{bmatrix} 0 \\ 0 \\ 1 \\ -1 \end{bmatrix} \middle| s, t, u \in \mathbb{R} \right\}.$$

The vector $\mathbf{v} = \begin{bmatrix} x \\ y \\ z \\ w \end{bmatrix}$ is in $W^{\perp}$ if and only if it is orthogonal to each of the three vectors that generate W, so

that $x - w = 0, y - w = 0, z - w = 0, z \in \mathbb{R}$. Hence, a basis for the orthogonal complement of W is $\left\{ \begin{bmatrix} 1 \\ 1 \\ 1 \\ 1 \end{bmatrix} \right\}$.

17. The two vectors that span W are linearly independent and orthogonal, so that an orthogonal basis for W is $B = \left\{ \begin{bmatrix} 2 \\ 0 \\ 0 \end{bmatrix}, \begin{bmatrix} 0 \\ -1 \\ 0 \end{bmatrix} \right\}$. Then

$$\mathbf{proj}_W \mathbf{v} = \frac{\begin{bmatrix} 1 \\ 2 \\ -3 \end{bmatrix} \cdot \begin{bmatrix} 2 \\ 0 \\ 0 \end{bmatrix}}{\begin{bmatrix} 2 \\ 0 \\ 0 \end{bmatrix} \cdot \begin{bmatrix} 2 \\ 0 \\ 0 \end{bmatrix}} \begin{bmatrix} 2 \\ 0 \\ 0 \end{bmatrix} + \frac{\begin{bmatrix} 1 \\ 2 \\ -3 \end{bmatrix} \cdot \begin{bmatrix} 0 \\ -1 \\ 0 \end{bmatrix}}{\begin{bmatrix} 0 \\ -1 \\ 0 \end{bmatrix} \cdot \begin{bmatrix} 0 \\ -1 \\ 0 \end{bmatrix}} \begin{bmatrix} 0 \\ -1 \\ 0 \end{bmatrix}$$

$$= \frac{2}{4} \begin{bmatrix} 2 \\ 0 \\ 0 \end{bmatrix} + \frac{-2}{1} \begin{bmatrix} 0 \\ -1 \\ 0 \end{bmatrix} = \begin{bmatrix} 1 \\ 2 \\ 0 \end{bmatrix}.$$

19. The spanning vectors for W are linearly independent but are not orthogonal. Using the Gram-Schmidt precess an orthogonal basis for W consists of the two vectors $\begin{bmatrix} 1 \\ 2 \\ 1 \end{bmatrix}$ and $\frac{1}{6} \begin{bmatrix} -13 \\ 4 \\ 5 \end{bmatrix}$. But we can also use the orthogonal basis consisting of the two vectors $\begin{bmatrix} 1 \\ 2 \\ 1 \end{bmatrix}$ and $\begin{bmatrix} -13 \\ 4 \\ 5 \end{bmatrix}$. Then

$$\mathbf{proj}_W \mathbf{v} = \frac{\begin{bmatrix} 1 \\ 2 \\ 1 \end{bmatrix} \cdot \begin{bmatrix} 1 \\ -3 \\ 5 \end{bmatrix}}{\begin{bmatrix} 1 \\ 2 \\ 1 \end{bmatrix} \cdot \begin{bmatrix} 1 \\ 2 \\ 1 \end{bmatrix}} \begin{bmatrix} 1 \\ 2 \\ 1 \end{bmatrix} + \frac{\begin{bmatrix} -13 \\ 4 \\ 5 \end{bmatrix} \cdot \begin{bmatrix} 1 \\ -3 \\ 5 \end{bmatrix}}{\begin{bmatrix} -13 \\ 4 \\ 5 \end{bmatrix} \cdot \begin{bmatrix} -13 \\ 4 \\ 5 \end{bmatrix}} \begin{bmatrix} -13 \\ 4 \\ 5 \end{bmatrix} = \begin{bmatrix} 0 \\ 0 \\ 0 \end{bmatrix}.$$

21. The spanning vectors for W are linearly independent but are not orthogonal. Using the Gram-Schmidt precess an orthogonal basis for W is $B = \left\{ \begin{bmatrix} 3 \\ 0 \\ -1 \\ 2 \end{bmatrix}, \begin{bmatrix} -5 \\ 21 \\ -3 \\ 6 \end{bmatrix} \right\}$. Then

$$\mathbf{proj}_W \mathbf{v} = \frac{4}{73} \begin{bmatrix} -5 \\ 21 \\ -3 \\ 6 \end{bmatrix}.$$

23. a. $W^\perp = \mathbf{span} \left\{ \begin{bmatrix} 1 \\ 3 \end{bmatrix} \right\}$ **b.** $\mathbf{proj}_W \mathbf{v} = \frac{1}{10} \begin{bmatrix} -3 \\ 1 \end{bmatrix}$ **c.** $\mathbf{u} = \mathbf{v} - \mathbf{proj}_W \mathbf{v} = \frac{1}{10} \begin{bmatrix} 3 \\ 9 \end{bmatrix}$

d. $\frac{1}{10} \begin{bmatrix} 3 \\ 9 \end{bmatrix} \cdot \begin{bmatrix} -3 \\ 1 \end{bmatrix} = 0$ **e.**

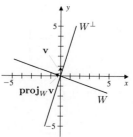

25. Notice that the vectors $\begin{bmatrix} 1 \\ 1 \\ -1 \end{bmatrix}$ and $\begin{bmatrix} -1 \\ 2 \\ 4 \end{bmatrix}$ are not orthogonal. Using the Gram-Schmidt process an

orthogonal basis for W is $\left\{ \begin{bmatrix} 1 \\ 1 \\ -1 \end{bmatrix}, \begin{bmatrix} 0 \\ 3 \\ 3 \end{bmatrix} \right\}$. **a.** $W^\perp = \textbf{span} \left\{ \begin{bmatrix} 2 \\ -1 \\ 1 \end{bmatrix} \right\}$

b. $\textbf{proj}_W \mathbf{v} = \frac{1}{3} \begin{bmatrix} 2 \\ 5 \\ 1 \end{bmatrix}$ **c.** $\mathbf{u} = \mathbf{v} - \textbf{proj}_W \mathbf{v} = \frac{1}{3} \begin{bmatrix} 4 \\ -2 \\ 2 \end{bmatrix}$

d. Since $\mathbf{u}$ is a scalar multiple of $\begin{bmatrix} 2 \\ -1 \\ 1 \end{bmatrix}$, **e.**

then $\mathbf{u}$ is in $W^\perp$.

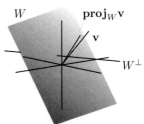

27. Let $\mathbf{w} \in W_2^\perp$, so $\langle \mathbf{w}, \mathbf{u} \rangle = 0$ for all $\mathbf{u} \in W_2$. Since $W_1 \subseteq W_2$, then $\langle \mathbf{w}, \mathbf{u} \rangle = 0$ for all $\mathbf{u} \in W_1$. Hence $\mathbf{w} \in W_1^\perp$, so $W_2^\perp \subseteq W_1^\perp$.

29. a. Let $A = \begin{bmatrix} d & e \\ f & g \end{bmatrix}$ and $B = \begin{bmatrix} a & b \\ b & c \end{bmatrix}$ be a matrix in W. Then

$$\langle A, B \rangle = \textbf{tr}(B^t A) = \textbf{tr}\left(\begin{bmatrix} a & b \\ b & c \end{bmatrix} \begin{bmatrix} d & e \\ f & g \end{bmatrix} \right) = \textbf{tr} \begin{bmatrix} ad + bf & ae + bg \\ bd + cf & be + cg \end{bmatrix}.$$

So $A \in W^\perp$ if and only if $ad + bf + be + cg = 0$ for all real numbers a, b, and c. This implies $A = \begin{bmatrix} 0 & e \\ -e & 0 \end{bmatrix}$. That is A is skew-symmetric.

b.
$$\begin{bmatrix} a & b \\ c & d \end{bmatrix} = \begin{bmatrix} a & \frac{b+c}{2} \\ \frac{b+c}{2} & d \end{bmatrix} + \begin{bmatrix} 0 & \frac{b-c}{2} \\ -\frac{b-c}{2} & 0 \end{bmatrix}$$

31. Let $\mathbf{w_0}$ be in W. Since $W^\perp = \{ \mathbf{v} \in W \mid \langle \mathbf{v}, \mathbf{w} \rangle = 0 \text{ for all } \mathbf{w} \in W \}$, then $\langle \mathbf{w_0}, \mathbf{v} \rangle = 0$, for every vector $\mathbf{v}$ in $W^\perp$. Hence, $\mathbf{w_0}$ is orthogonal to every vector in $W^\perp$, so $\mathbf{w_0} \in (W^\perp)^\perp$. That is, $W \subseteq (W^\perp)^\perp$.

Now let $\mathbf{w_0} \in (W^\perp)^\perp$. Since $V = W \oplus W^\perp$, then $\mathbf{w_0} = \mathbf{w} + \mathbf{v}$, where $\mathbf{w} \in W$ and $\mathbf{v} \in W^\perp$. So $\langle \mathbf{v}, \mathbf{w_0} \rangle = 0$, since $\mathbf{w_0} \in (W^\perp)^\perp$ and $\mathbf{v} \in W^\perp$ and $\langle \mathbf{v}, \mathbf{w} \rangle = 0$, since $\mathbf{w} \in W$ and $\mathbf{v} \in W^\perp$. Then

$$0 = \langle \mathbf{v}, \mathbf{w_0} \rangle = \langle \mathbf{v}, \mathbf{w} + \mathbf{v} \rangle = \langle \mathbf{v}, \mathbf{w} \rangle + \langle \mathbf{v}, \mathbf{v} \rangle = \langle \mathbf{v}, \mathbf{v} \rangle.$$

Therefore, since V is an inner product space $\mathbf{v} = \mathbf{0}$, so $\mathbf{w_0} = \mathbf{w} \in W$. Hence, $(W^\perp)^\perp \subseteq W$. Since both containments hold, we have $W = (W^\perp)^\perp$.

Exercise Set 6.5

1. a. To find the least squares solution it is equivalent to solving the normal equation

$A^t A \begin{bmatrix} x \\ y \end{bmatrix} = A^t \begin{bmatrix} 4 \\ 1 \\ 5 \end{bmatrix}$. That is,

$$\begin{bmatrix} 1 & 1 & 2 \\ 3 & 3 & 3 \end{bmatrix} \begin{bmatrix} 1 & 3 \\ 1 & 3 \\ 2 & 3 \end{bmatrix} \begin{bmatrix} x \\ y \end{bmatrix} = \begin{bmatrix} 1 & 1 & 2 \\ 3 & 3 & 3 \end{bmatrix} \begin{bmatrix} 4 \\ 1 \\ 5 \end{bmatrix} \Leftrightarrow \begin{bmatrix} 6 & 12 \\ 12 & 27 \end{bmatrix} \begin{bmatrix} x \\ y \end{bmatrix} = \begin{bmatrix} 15 \\ 30 \end{bmatrix}.$$

Hence, the least squares solution is $\widehat{\mathbf{x}} = \begin{bmatrix} 5/2 \\ 0 \end{bmatrix}$. **b.** Since the orthogonal projection of $\mathbf{b}$ onto W is $A\widehat{\mathbf{x}}$, we

have that $\mathbf{w_1} = A\widehat{\mathbf{x}} = \begin{bmatrix} 5/2 \\ 5/2 \\ 5 \end{bmatrix}$, and $\mathbf{w_2} = \mathbf{b} - \mathbf{w_1} = \begin{bmatrix} 3/2 \\ -3/2 \\ 0 \end{bmatrix}$.

3. Let

$$A = \begin{bmatrix} 1965 & 1 \\ 1970 & 1 \\ 1975 & 1 \\ 1980 & 1 \\ 1985 & 1 \\ 1990 & 1 \\ 1995 & 1 \\ 2000 & 1 \\ 2004 & 1 \end{bmatrix} \quad \text{and} \quad \mathbf{b} = \begin{bmatrix} 927 \\ 1187 \\ 1449 \\ 1710 \\ 2004 \\ 2185 \\ 2513 \\ 2713 \\ 2803 \end{bmatrix}.$$

a. The figure shows the scatter plot of the data, which approximates a linear increasing trend. Also shown in the figure is the best fit line found in part (b).

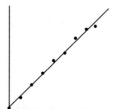

b. The least squares solution is given by the solution to the normal equation

$A^t A \begin{bmatrix} m \\ b \end{bmatrix} = A^t \mathbf{b}$, which is equivalent to

$$\begin{bmatrix} 35459516 & 17864 \\ 17864 & 9 \end{bmatrix} \begin{bmatrix} m \\ b \end{bmatrix} = \begin{bmatrix} 34790257 \\ 17491 \end{bmatrix}.$$

The solution to the system gives the line that best fits the data, $y = \frac{653089}{13148}x - \frac{317689173}{3287}$.

5. a. The figure shows the scatter plot of the data, which approximates a linear increasing trend. Also shown in the figure is the best fit line found in part (b).

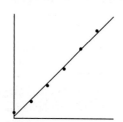

b. The line that best fits the data is

$$y = 0.07162857143x - 137.2780952.$$

7. a. The Fourier polynomials are

$$p_2(x) = 2\sin(x) - \sin(2x)$$

$$p_3(x) = 2\sin(x) - \sin(2x) + \frac{2}{3}\sin(3x)$$

$$p_4(x) = 2\sin(x) - \sin(2x) + \frac{2}{3}\sin(3x)$$
$$- \frac{1}{2}\sin(4x)$$

$$p_5(x) = 2\sin(x) - \sin(2x) + \frac{2}{3}\sin(3x)$$
$$- \frac{1}{2}\sin(4x) + \frac{2}{5}\sin(5x).$$

b. The graph of the function $f(x) = x$, on the interval $-\pi \le x \le \pi$ and the Fourier approximations are shown in the figure.

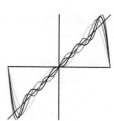

9. a. The Fourier polynomials are

$$p_2(x) = \frac{1}{3}\pi^2 - 4\cos(x) + \cos(2x)$$

$$p_3(x) = \frac{1}{3}\pi^2 - 4\cos(x) + \cos(2x) - \frac{4}{9}\cos(3x)$$

$$p_4(x) = \frac{1}{3}\pi^2 - 4\cos(x) + \cos(2x) - \frac{4}{9}\cos(3x)$$
$$+ \frac{1}{4}\cos(4x)$$

$$p_5(x) = \frac{1}{3}\pi^2 - 4\cos(x) + \cos(2x) - \frac{4}{9}\cos(3x)$$
$$+ \frac{1}{4}\cos(4x) - \frac{4}{25}\cos(5x).$$

b. The graph of the function $f(x) = x^2$, on the interval $-\pi \le x \le \pi$ and the Fourier approximations are shown in the figure.

Exercise Set 6.6

Diagonalization of matrices was considered earlier and several criteria were given for determining when a matrix can be factored in this specific form. An $n \times n$ matrix is diagonalizable if and only if it has n linearly independent eigenvectors. Also if A has n distinct eigenvalues, then A is diagonalizable. If A is an $n \times n$ real symmetric matrix, then the eigenvalues are all real numbers and the eigenvectors corresponding to distinct eigenvalues are orthogonal. We also have that every real symmetric matrix has an orthogonal diagonalization. That is, there is an orthogonal matrix P, so that $P^{-1} = P^t$, and a diagonal matrix D such that $A = PDP^{-1} = PDP^t$. The column vectors of an orthogonal matrix form an orthonormal basis for $\mathbb{R}^n$. If the symmetric matrix A has an eigenvalue of geometric multiplicity greater than 1, then the corresponding linearly independent eigenvectors that generate the eigenspace may not be orthogonal. So a process for finding an orthogonal diagonalization of a real $n \times n$ symmetric matrix A is:

- Find the eigenvalues and corresponding eigenvectors of A.

- Since A is diagonalizable there are n linearly independent eigenvectors. If necessary use the Gram-Schmidt process to find an orthonormal set of eigenvectors.

- Form the orthogonal matrix P with column vectors determined in the previous step.

- Form the diagonal matrix with diagonal entries the eigenvalues of A. The eigenvalues are placed on the diagonal in the same order as the eigenvectors are used to define P. If an eigenvalue has algebraic multiplicity m, then the corresponding eigenvalue and eigenvector are repeated in D and P m-times, respectively.

- The matrix A has the factorization $A = PDP^{-1} = PDP^t$.

■ Solutions to Odd Exercises

1. The eigenvalues are the solutions to the characteristic equation $\det(A - \lambda I) = 0$, that is, $\lambda_1 = 3$ and $\lambda_2 = -1$

3. $\lambda_1 = 1, \lambda_2 = -3, \lambda_3 = 3$.

5. Since the eigenvalues are $\lambda_1 = -3$ with eigenvector $\mathbf{v_1} = \begin{bmatrix} -1 \\ 2 \end{bmatrix}$ and $\lambda_2 = 2$ with eigenvector $\mathbf{v_2} = \begin{bmatrix} 2 \\ 1 \end{bmatrix}$, then $\mathbf{v_1} \cdot \mathbf{v_2} = 0$, so the eigenvectors are orthogonal.

7. Since the eigenvalues are $\lambda_1 = 1$ with eigenvector $\mathbf{v_1} = \begin{bmatrix} 1 \\ 0 \\ 1 \end{bmatrix}$, $\lambda_2 = -3$ with eigenvector $\mathbf{v_2} = \begin{bmatrix} -1 \\ 2 \\ 1 \end{bmatrix}$,

and $\lambda_3 = 3$ with eigenvector $\mathbf{v_3} = \begin{bmatrix} -1 \\ -1 \\ 1 \end{bmatrix}$, then $\mathbf{v_1} \cdot \mathbf{v_2} = \mathbf{v_1} \cdot \mathbf{v_3} = \mathbf{v_2} \cdot \mathbf{v_3} = 0$, so the eigenvectors are

pairwise orthogonal.

9. The eigenvalues of A are $\lambda_1 = 3$ and $\lambda_2 = -1$ with multiplicity 2. Moreover, there are three linearly

independent eigenvectors and the eigenspaces are $V_3 = \text{span} \left\{ \begin{bmatrix} 1 \\ 0 \\ 1 \end{bmatrix} \right\}$ and $V_{-1} = \text{span} \left\{ \begin{bmatrix} -1 \\ 0 \\ 1 \end{bmatrix}, \begin{bmatrix} 0 \\ 1 \\ 0 \end{bmatrix} \right\}$.

Consequently, $\dim(V_3) + \dim(V_{-1}) = 1 + 2 = 3$.

11. The eigenspaces are $V_3 = \text{span} \left\{ \begin{bmatrix} 3 \\ 1 \\ 1 \\ 1 \end{bmatrix} \right\}$, $V_{-3} = \text{span} \left\{ \begin{bmatrix} 0 \\ -2 \\ 1 \\ 1 \end{bmatrix} \right\}$, and

$V_{-1} = \text{span} \left\{ \begin{bmatrix} -1 \\ 1 \\ 2 \\ 0 \end{bmatrix}, \begin{bmatrix} 0 \\ 0 \\ -1 \\ 1 \end{bmatrix} \right\}$, so $\dim(V_3) + \dim(V_{-3}) + \dim(V_{-1}) = 1 + 1 + 2 = 4$.

13. Since $A^t A = \begin{bmatrix} \sqrt{3}/2 & 1/2 \\ -1/2 & \sqrt{3}/2 \end{bmatrix} \begin{bmatrix} \sqrt{3}/2 & -1/2 \\ 1/2 & \sqrt{3}/2 \end{bmatrix} = \begin{bmatrix} 1 & 0 \\ 0 & 1 \end{bmatrix}$, the inverse of A is A^t, so the matrix A is

orthogonal.

15. Since $A^t A = \begin{bmatrix} \sqrt{2}/2 & \sqrt{2}/2 & 0 \\ -\sqrt{2}/2 & \sqrt{2}/2 & 0 \\ 0 & 0 & 1 \end{bmatrix} \begin{bmatrix} \sqrt{2}/2 & -\sqrt{2}/2 & 0 \\ \sqrt{2}/2 & \sqrt{2}/2 & 0 \\ 0 & 0 & 1 \end{bmatrix} = \begin{bmatrix} 1 & 0 & 0 \\ 0 & 1 & 0 \\ 0 & 0 & 1 \end{bmatrix}$, the inverse of A is A^t, so

the matrix A is orthogonal.

17. The eigenvalues of the matrix A are -1 and 7 with corresponding orthogonal eigenvectors $\begin{bmatrix} -1 \\ 1 \end{bmatrix}$

and $\begin{bmatrix} 1 \\ 1 \end{bmatrix}$, respectively. An orthonormal pair of eigenvectors is $\frac{1}{\sqrt{2}} \begin{bmatrix} -1 \\ 1 \end{bmatrix}$ and $\frac{1}{\sqrt{2}} \begin{bmatrix} 1 \\ 1 \end{bmatrix}$. Let $P =$

$\begin{bmatrix} -1/\sqrt{2} & 1/\sqrt{2} \\ 1/\sqrt{2} & 1/\sqrt{2} \end{bmatrix}$, so that $P^{-1} A P = D = \begin{bmatrix} -1 & 0 \\ 0 & 7 \end{bmatrix}$.

19. If $P = \begin{bmatrix} -1/\sqrt{2} & 1/\sqrt{2} \\ 1/\sqrt{2} & 1/\sqrt{2} \end{bmatrix}$, then $P^{-1} A P = D = \begin{bmatrix} -4 & 0 \\ 0 & 2 \end{bmatrix}$.

21. The eigenvalues of A are $-2, 2$, and 1 with corresponding eigenvectors $\begin{bmatrix} -1 \\ -2 \\ 1 \end{bmatrix}, \begin{bmatrix} 1 \\ 0 \\ 1 \end{bmatrix}$, and $\begin{bmatrix} -1 \\ 1 \\ 1 \end{bmatrix}$,

respectively. Since the eigenvectors are pairwise orthogonal, let P be the matrix with column vectors unit

eigenvectors. That is, let $P = \begin{bmatrix} -1/\sqrt{3} & 1/\sqrt{2} & -1/\sqrt{6} \\ 1/\sqrt{3} & 0 & -2/\sqrt{6} \\ 1/\sqrt{3} & 1/\sqrt{2} & 1/\sqrt{6} \end{bmatrix}$, then $P^{-1} A P = D = \begin{bmatrix} 1 & 0 & 0 \\ 0 & 2 & 0 \\ 0 & 0 & -2 \end{bmatrix}$.

23. Since A and B are orthogonal matrices, then $AA^t = BB^t = I$, so that

$$(AB)(AB)^t = AB(B^t A^t) = A(BB^t)A^t = AIA^t = AA^t = I.$$

Since the inverse of AB is $(AB)^t$, then AB is an orthogonal matrix. Similarly, $(BA)(BA)^t = I$.

25. We need to show that the inverse of A^t is $(A^t)^t = A$. Since A is orthogonal, then $AA^t = I$, that is, A is
the inverse of A^t and hence, A^t is also orthogonal.

27. a. Since $\cos^2 \theta + \sin^2 \theta = 1$, then

$$A^t A = \begin{bmatrix} \cos\theta & -\sin\theta \\ \sin\theta & \cos\theta \end{bmatrix} \begin{bmatrix} \cos\theta & \sin\theta \\ -\sin\theta & \cos\theta \end{bmatrix} = \begin{bmatrix} 1 & 0 \\ 0 & 1 \end{bmatrix},$$

so the matrix A is orthogonal.

b. Let $A = \begin{bmatrix} a & b \\ c & d \end{bmatrix}$. Since A is orthogonal, then

$$\begin{bmatrix} a & b \\ c & d \end{bmatrix} \begin{bmatrix} a & c \\ b & d \end{bmatrix} = \begin{bmatrix} a^2 + b^2 & ac + bd \\ ac + bd & c^2 + d^2 \end{bmatrix} = \begin{bmatrix} 1 & 0 \\ 0 & 1 \end{bmatrix} \Leftrightarrow \begin{cases} a^2 + b^2 & = 1 \\ ac + bd & = 0 \\ c^2 + d^2 & = 1 \end{cases}.$$

Let $\mathbf{v_1} = \begin{bmatrix} a \\ b \end{bmatrix}$ and let θ be the angle that $\mathbf{v_1}$ makes with the horizontal axis. Since $a^2 + b^2 = 1$, then $\mathbf{v_1}$ is a unit vector. Therefore, $a = \cos\theta$ and $b = \sin\theta$. Now let $\mathbf{v_2} = \begin{bmatrix} c \\ d \end{bmatrix}$. Since $ac + bd = 0$ then $\mathbf{v_1}$ and $\mathbf{v_2}$ are orthogonal. There are two cases.

Case 1. $c = \cos(\theta + \pi/2) = -\sin\theta$ and $d = \sin(\theta + \pi/2) = \cos\theta$, so that $A = \begin{bmatrix} \cos\theta & \sin\theta \\ -\sin\theta & \cos\theta \end{bmatrix}$.

Case 2. $c = \cos(\theta - \pi/2) = \sin\theta$ and $d = \sin(\theta - \pi/2) = -\cos\theta$, so that $A = \begin{bmatrix} \cos\theta & \sin\theta \\ \sin\theta & -\cos\theta \end{bmatrix}$.

c. If $\det(A) = 1$, then by part (b), $T(\mathbf{v}) = A\mathbf{v}$ with $A = \begin{bmatrix} \cos\theta & \sin\theta \\ -\sin\theta & \cos\theta \end{bmatrix}$. Therefore

$$T(\mathbf{v}) = \begin{bmatrix} \cos\theta & \sin\theta \\ -\sin\theta & \cos\theta \end{bmatrix} \begin{bmatrix} x \\ y \end{bmatrix} = \begin{bmatrix} \cos(-\theta) & -\sin(-\theta) \\ \sin(-\theta) & \cos(-\theta) \end{bmatrix} \begin{bmatrix} x \\ y \end{bmatrix},$$

which is a rotation of a vector by $-\theta$ radians. If $\det(A) = -1$, then by part (b), $T(\mathbf{v}) = A'\mathbf{v}$ with $A' = \begin{bmatrix} \cos\theta & \sin\theta \\ \sin\theta & -\cos\theta \end{bmatrix}$. Observe that

$$A' = \begin{bmatrix} \cos\theta & -\sin\theta \\ \sin\theta & \cos\theta \end{bmatrix} \begin{bmatrix} 1 & 0 \\ 0 & -1 \end{bmatrix}.$$

Hence, in this case, T is a reflection through the x-axis followed by a rotation through the angle θ.

29. Suppose $D = P^t AP$, where P is an orthogonal matrix, that is $P^{-1} = P^t$. Then

$$D^t = (P^t AP)^t = P^t A^t P.$$

Since D is a diagonal matrix then $D^t = D$, so we also have $D = P^t A^t P$ and hence, $P^t AP = P^t A^t P$. Then $P(P^t AP)P^t = P(P^t A^t P)P^t$. Since $PP^t = I$, we have that $A = A^t$, and hence, the matrix A is symmetric.

31. a. If $\mathbf{v} = \begin{bmatrix} v_1 \\ v_2 \\ \vdots \\ v_n \end{bmatrix}$, then $\mathbf{v}^t \mathbf{v} = v_1^2 + \ldots + v_n^2$.

b. Consider the equation $A\mathbf{v} = \lambda\mathbf{v}$. Now take the transpose of both sides to obtain $\mathbf{v}^t A^t = \lambda\mathbf{v}^t$. Since A is skew symmetric this is equivalent to

$$\mathbf{v}^t(-A) = \lambda\mathbf{v}^t.$$

Now, right multiplication of both sides by $\mathbf{v}$ gives $\mathbf{v}^t(-A\mathbf{v}) = \lambda\mathbf{v}^t\mathbf{v}$ or equivalently, $\mathbf{v}^t(-\lambda\mathbf{v}) = \lambda\mathbf{v}^t\mathbf{v}$. Hence, $2\lambda\mathbf{v}^t\mathbf{v} = 0$, so that by part (a),

$$2\lambda(v_1^2 + \ldots + v_n^2) = 0, \quad \text{that is } \lambda = 0 \text{ or } \mathbf{v} = \mathbf{0}.$$

Since $\mathbf{v}$ is an eigenvector $\mathbf{v} \neq \mathbf{0}$, and hence $\lambda = 0$. Therefore, the only eigenvalue of A is $\lambda = 0$.

Exercise Set 6.7

1. Let $\mathbf{x} = \begin{bmatrix} x \\ y \end{bmatrix}$, $A = \begin{bmatrix} 27 & -9 \\ -9 & 3 \end{bmatrix}$, and $\mathbf{b} = \begin{bmatrix} 1 \\ 3 \end{bmatrix}$. Then the quadratic equation is equivalent to $\mathbf{x}^t A \mathbf{x} + \mathbf{b}^t \mathbf{x} = 0$. The next step is to diagonalize the matrix A. The eigenvalues of A are 30 and 0 with corresponding eigenvectors $\begin{bmatrix} -3 \\ 1 \end{bmatrix}$ and $\begin{bmatrix} 1 \\ 3 \end{bmatrix}$, respectively. Notice that the eigenvectors are not orthogonal. Using the Gram-Schmidt process unit orthogonal vectors are $\mathbf{v_1} = \begin{bmatrix} -\frac{3\sqrt{10}}{10} \\ \frac{\sqrt{10}}{10} \end{bmatrix}$ and $\mathbf{v_2} = \begin{bmatrix} \frac{\sqrt{10}}{10} \\ \frac{3\sqrt{10}}{10} \end{bmatrix}$. The matrix with column vectors $\mathbf{v_1}$ and $\mathbf{v_2}$ is orthogonal, but the determinant is -1. By interchanging the column vectors the resulting matrix is orthogonal and is a rotation. Let

$$P = \begin{bmatrix} \frac{\sqrt{10}}{10} & -\frac{3\sqrt{10}}{10} \\ \frac{3\sqrt{10}}{10} & \frac{\sqrt{10}}{10} \end{bmatrix} \quad \text{and} \quad D = \begin{bmatrix} 0 & 0 \\ 0 & 30 \end{bmatrix},$$

so that the equation is transformed to

$$(\mathbf{x}')^t D\mathbf{x}' + \mathbf{b}^t P\mathbf{x}' = 0, \quad \text{that is} \quad 30(y')^2 + \sqrt{10}x' = 0.$$

3. Let $\mathbf{x} = \begin{bmatrix} x \\ y \end{bmatrix}$, $A = \begin{bmatrix} 12 & 4 \\ 4 & 12 \end{bmatrix}$, and $\mathbf{b} = \begin{bmatrix} 0 \\ 0 \end{bmatrix}$. The matrix form of the quadratic equation is $\mathbf{x}^t A \mathbf{x} - 8 = 0$. The eigenvalues of A are 16 and 8 with orthogonal eigenvectors $\begin{bmatrix} 1 \\ 1 \end{bmatrix}$ and $\begin{bmatrix} -1 \\ 1 \end{bmatrix}$. Orthonormal eigenvectors are $\frac{\sqrt{2}}{2}\begin{bmatrix} 1 \\ 1 \end{bmatrix}$ and $\frac{\sqrt{2}}{2}\begin{bmatrix} -1 \\ 1 \end{bmatrix}$, so that $P = \begin{bmatrix} \frac{\sqrt{2}}{2} & -\frac{\sqrt{2}}{2} \\ \frac{\sqrt{2}}{2} & \frac{\sqrt{2}}{2} \end{bmatrix}$ is an orthogonal matrix whose action on a vector is a rotation. If $D = \begin{bmatrix} 16 & 0 \\ 0 & 8 \end{bmatrix}$, then the quadratic equation is transformed to $(\mathbf{x}')^t D\mathbf{x}' - 8 = 0$, that is, $2(x')^2 + (y')^2 = 1$.

5. The transformed quadratic equation is $\frac{(x')^2}{2} - \frac{(y')^2}{4} = 1$.

7. a. $[x \ y]\begin{bmatrix} 4 & 0 \\ 0 & 16 \end{bmatrix}\begin{bmatrix} x \\ y \end{bmatrix} - 16 = 0$

b. The action of the matrix

$$P = \begin{bmatrix} \cos\frac{\pi}{4} & -\sin\frac{\pi}{4} \\ \sin\frac{\pi}{4} & \cos\frac{\pi}{4} \end{bmatrix} = \begin{bmatrix} \frac{\sqrt{2}}{2} & -\frac{\sqrt{2}}{2} \\ \frac{\sqrt{2}}{2} & \frac{\sqrt{2}}{2} \end{bmatrix}$$

on a vector is a counter clockwise rotation of $45°$. Then $P\begin{bmatrix} 4 & 0 \\ 0 & 16 \end{bmatrix}P^t = \begin{bmatrix} 10 & -6 \\ -6 & 10 \end{bmatrix}$, so the quadratic equation that describes the original conic rotated $45°$ is

$$[x y]\begin{bmatrix} 10 & -6 \\ -6 & 10 \end{bmatrix}\begin{bmatrix} x \\ y \end{bmatrix} - 16 = 0, \quad \text{that is } 10x^2 - 12xy + 10y^2 - 16 = 0.$$

9. a. $7x^2 + 6\sqrt{3}xy + 13y^2 - 16 = 0$ **b.** $7(x-3)^2 + 6\sqrt{3}(x-3)(y-2) + 13(y-2)^2 - 16 = 0$

Exercise Set 6.8

1. The singular values of the matrix are $\sigma_1 = \sqrt{\lambda_1}, \sigma_2 = \sqrt{\lambda_2}$, where λ_1 and λ_2 are the eigenvalues of $A^t A$. Then $A^t A = \begin{bmatrix} -2 & 1 \\ -2 & 1 \end{bmatrix}\begin{bmatrix} -2 & -2 \\ 1 & 1 \end{bmatrix} = \begin{bmatrix} 5 & 5 \\ 5 & 5 \end{bmatrix}$, so $\sigma_1 = \sqrt{10}$ and $\sigma_2 = 0$.

3. Since

$$A^t A = \begin{bmatrix} 1 & 2 & -2 \\ 0 & -1 & 1 \\ 2 & -1 & 1 \end{bmatrix} \begin{bmatrix} 1 & 0 & 2 \\ 2 & -1 & -1 \\ -2 & 1 & 1 \end{bmatrix} = \begin{bmatrix} -9 & -4 & -2 \\ -4 & 2 & 2 \\ -2 & 2 & 6 \end{bmatrix}$$

has eigenvalues $0, 5$ and 12, then the singular values of A are $\sigma_1 = 2\sqrt{3}, \sigma_2 = \sqrt{5}$, and $\sigma_3 = 0$.

5. Step 1: The eigenvalues of $A^t A$ are 4 and 64 with corresponding eigenvectors $\begin{bmatrix} -1 \\ 1 \end{bmatrix}$ and $\begin{bmatrix} 1 \\ 1 \end{bmatrix}$, which

are orthogonal but are not orthonormal. An orthonormal pair of eigenvectors is $\mathbf{v_1} = \begin{bmatrix} \frac{1}{\sqrt{2}} \\ \frac{1}{\sqrt{2}} \end{bmatrix}$ and $\mathbf{v_2} =$

$\begin{bmatrix} \frac{1}{\sqrt{2}} \\ -\frac{1}{\sqrt{2}} \end{bmatrix}$. Let

$$V = \begin{bmatrix} \frac{1}{\sqrt{2}} & \frac{1}{\sqrt{2}} \\ \frac{1}{\sqrt{2}} & -\frac{1}{\sqrt{2}} \end{bmatrix}.$$

Step 2: The singular values of A are $\sigma_1 = \sqrt{64} = 8$ and $\sigma_2 = \sqrt{4} = 2$. Since the size of the matrix Σ is the same size as A, we have that

$$\Sigma = \begin{bmatrix} 8 & 0 \\ 0 & 2 \end{bmatrix}.$$

Step 3: The matrix U is defined by

$$U = \begin{bmatrix} \frac{1}{\sigma_1} A\mathbf{v_1} & \frac{1}{\sigma_2} A\mathbf{v_2} \end{bmatrix} = \begin{bmatrix} \frac{1}{\sqrt{2}} & \frac{1}{\sqrt{2}} \\ \frac{1}{\sqrt{2}} & -\frac{1}{\sqrt{2}} \end{bmatrix}.$$

Step 4: The SVD of A is

$$A = \begin{bmatrix} \frac{1}{\sqrt{2}} & \frac{1}{\sqrt{2}} \\ \frac{1}{\sqrt{2}} & -\frac{1}{\sqrt{2}} \end{bmatrix} \begin{bmatrix} 8 & 0 \\ 0 & 2 \end{bmatrix} \begin{bmatrix} \frac{1}{\sqrt{2}} & \frac{1}{\sqrt{2}} \\ \frac{1}{\sqrt{2}} & -\frac{1}{\sqrt{2}} \end{bmatrix}.$$

7. The SVD of A is

$$A = \begin{bmatrix} 0 & 1 \\ 1 & 0 \end{bmatrix} \begin{bmatrix} \sqrt{2} & 0 & 0 \\ 0 & 1 & 0 \end{bmatrix} \begin{bmatrix} 0 & \frac{1}{\sqrt{2}} & \frac{1}{\sqrt{2}} \\ 1 & 0 & 0 \\ 0 & -\frac{1}{\sqrt{2}} & \frac{1}{\sqrt{2}} \end{bmatrix}.$$

9. a. If $\mathbf{x} = \begin{bmatrix} x_1 \\ x_2 \end{bmatrix}$, then the solution to the linear system $A\mathbf{x} = \begin{bmatrix} 2 \\ 2 \end{bmatrix}$ is $x_1 = 2, x_2 = 0$ **b.** The solution

to the linear system $A\mathbf{x} = \begin{bmatrix} 2 \\ 2.000000001 \end{bmatrix}$ is $x_1 = 1, x_2 = 1$. **c.** The condition number of the matrix A

is $\frac{\sigma_1}{\sigma_2} \approx 6324555$, which is relatively large. Notice that a small change in the vector $\mathbf{b}$ in the linear system $A\mathbf{x} = \mathbf{b}$ results in a significant difference in the solutions.

Review Exercises Chapter 6

1. a. Since the set B contains three vectors in $\mathbb{R}^3$ it is sufficient to show that B is linearly independent. Since

$$\begin{bmatrix} 1 & 1 & 2 \\ 0 & 0 & 1 \\ 1 & 0 & 0 \end{bmatrix} \longrightarrow \begin{bmatrix} 1 & 0 & 0 \\ 0 & 1 & 0 \\ 0 & 0 & 1 \end{bmatrix},$$

the only solution to the homogeneous linear system $\begin{bmatrix} 1 & 1 & 2 \\ 0 & 0 & 1 \\ 1 & 0 & 0 \end{bmatrix} \begin{bmatrix} c_1 \\ c_2 \\ c_3 \end{bmatrix} = \begin{bmatrix} 0 \\ 0 \\ 0 \end{bmatrix}$ is the trivial solution, so B is linearly independent and hence, is a basis for $\mathbb{R}^3$. **b.** Notice that the vectors in B are not pairwise orthogonal. Using the Gram-Schmidt process an orthonormal basis is $\left\{ \begin{bmatrix} 0 \\ 1 \\ 0 \end{bmatrix}, \begin{bmatrix} \sqrt{2}/2 \\ 0 \\ -\sqrt{2}/2 \end{bmatrix}, \begin{bmatrix} \sqrt{2}/2 \\ 0 \\ \sqrt{2}/2 \end{bmatrix} \right\}$. **c.**

Again the spanning vectors for W are not orthogonal, which is required to find $\mathbf{proj}_W \mathbf{v}$. The Gram-Schmidt process yields the orthogonal vectors $\begin{bmatrix} 1 \\ 0 \\ 1 \end{bmatrix}$ and $\begin{bmatrix} 1/2 \\ 0 \\ -1/2 \end{bmatrix}$, which also span W. Then

$$\mathbf{proj}_W \mathbf{v} = \frac{\begin{bmatrix} -2 \\ 1 \\ -1 \end{bmatrix} \cdot \begin{bmatrix} 1 \\ 0 \\ 1 \end{bmatrix}}{\begin{bmatrix} 1 \\ 0 \\ 1 \end{bmatrix} \cdot \begin{bmatrix} 1 \\ 0 \\ 1 \end{bmatrix}} \begin{bmatrix} 1 \\ 0 \\ 1 \end{bmatrix} + \frac{\begin{bmatrix} -2 \\ 1 \\ -1 \end{bmatrix} \cdot \begin{bmatrix} 1/2 \\ 0 \\ -1/2 \end{bmatrix}}{\begin{bmatrix} 1/2 \\ 0 \\ -1/2 \end{bmatrix} \cdot \begin{bmatrix} 1/2 \\ 0 \\ -1/2 \end{bmatrix}} \begin{bmatrix} 1/2 \\ 0 \\ -1/2 \end{bmatrix} = \begin{bmatrix} -2 \\ 0 \\ -1 \end{bmatrix}.$$

3. a. If $\begin{bmatrix} x \\ y \\ z \end{bmatrix} \in W$, then $\begin{bmatrix} x \\ y \\ z \end{bmatrix} \cdot \begin{bmatrix} a \\ b \\ c \end{bmatrix} = ax + by + cz = 0$, so $\begin{bmatrix} a \\ b \\ c \end{bmatrix}$ is in $W^\perp$. **b.** The orthogonal complement of W is the set of all vectors that are orthogonal to everything in W, that is, $W^\perp = \mathbf{span} \left\{ \begin{bmatrix} a \\ b \\ c \end{bmatrix} \right\}$. So $W^\perp$ is

the line in the direction of $\begin{bmatrix} a \\ b \\ c \end{bmatrix}$ and which is perpendicular (the normal vector) to the plane $ax + by + cz = 0$.

c. $\mathbf{proj}_{W^\perp} \mathbf{v} = \dfrac{\begin{bmatrix} a \\ b \\ c \end{bmatrix} \cdot \begin{bmatrix} x_1 \\ x_2 \\ x_3 \end{bmatrix}}{\begin{bmatrix} a \\ b \\ c \end{bmatrix} \cdot \begin{bmatrix} a \\ b \\ c \end{bmatrix}} \begin{bmatrix} a \\ b \\ c \end{bmatrix} = \dfrac{ax_1 + bx_2 + cx_3}{a^2 + b^2 + c^2} \begin{bmatrix} a \\ b \\ c \end{bmatrix}$ **d.** $\| \mathbf{proj}_{W^\perp} \mathbf{v} \| = \dfrac{|ax_1 + bx_2 + cx_3|}{\sqrt{a^2 + b^2 + c^2}}$

Note that this norm is the distance from the point (x_1, x_2, x_3) to the plane.

5. a. Since $\langle 1, \cos x \rangle = \int_{-\pi}^{\pi} \cos x \, dx = 0$, $\langle 1, \sin x \rangle = \int_{-\pi}^{\pi} \sin x \, dx = 0$, and $\langle \cos x, \sin x \rangle = \int_{-\pi}^{\pi} \cos x \sin x \, dx = 0$, the set W is orthogonal. **b.** Since the polynomials in W are orthogonal an orthonormal basis is obtained by normalizing the polynomials. For example, $\|1\| = \sqrt{\int_{-\pi}^{\pi} dx} = \sqrt{2\pi}$. Similarly, $\|\cos x\| = \sqrt{\pi}$, and $\|\sin x\| = \sqrt{\pi}$, so an orhtonormal basis for W is $\left\{ \frac{1}{\sqrt{2\pi}}, \frac{1}{\sqrt{\pi}} \cos x, \frac{1}{\sqrt{\pi}} \sin x \right\}$.

c. $\mathbf{proj}_W x^2 = \left\langle \frac{1}{\sqrt{2\pi}}, x^2 \right\rangle \frac{1}{\sqrt{2\pi}} + \left\langle \frac{1}{\sqrt{\pi}} \cos x, x^2 \right\rangle \frac{1}{\sqrt{\pi}} \cos x + \left\langle \frac{1}{\sqrt{\pi}} \sin x, x^2 \right\rangle \frac{1}{\sqrt{\pi}} \sin x = \frac{1}{3}\pi^2 - 4\cos x$
d. $\| \mathbf{proj}_W x^2 \| = \frac{1}{3}\sqrt{2\pi^5 + 144\pi}$

7. Let $B = \{\mathbf{v_1}, \mathbf{v_2}, \ldots, \mathbf{v_n}\}$ be an orthonormal basis and $[\mathbf{v}]_B = \begin{bmatrix} v_1 \\ v_2 \\ \vdots \\ v_3 \end{bmatrix}$. Then there are scalars $c_1, \ldots, c_n$

such that $\mathbf{v} = c_1 \mathbf{v_1} + c_2 \mathbf{v_2} + \cdots + c_n \mathbf{v_n}$. Using the properties of an inner product and the fact that the vectors are orthonormal,

$$\| \mathbf{v} \| = \sqrt{\langle \mathbf{v}, \mathbf{v} \rangle} = \sqrt{\mathbf{v} \cdot \mathbf{v}} = \sqrt{\langle c_1 \mathbf{v_1} + c_2 \mathbf{v_2} + \cdots + c_n \mathbf{v_n}, c_1 \mathbf{v_1} + c_2 \mathbf{v_2} + \cdots + c_n \mathbf{v_n} \rangle}$$

$$= \sqrt{\langle c_1 \mathbf{v_1}, c_1 \mathbf{v_1} \rangle + \cdots + \langle c_n \mathbf{v_n}, c_n \mathbf{v_n} \rangle} = \sqrt{c_1^2 \langle \mathbf{v_1}, \mathbf{v_1} \rangle + \cdots + c_n^2 \langle \mathbf{v_n}, \mathbf{v_n} \rangle}$$

$$= \sqrt{c_1^2 + \cdots + c_n^2}.$$

If the basis is orthogonal, then $\| \mathbf{v} \| = \sqrt{c_1^2 \langle \mathbf{v_1}, \mathbf{v_1} \rangle + \cdots + c_n^2 \langle \mathbf{v_n}, \mathbf{v_n} \rangle}$.

9. a. Since $A = \begin{bmatrix} 1 & 0 & -1 \\ 1 & -1 & 2 \\ 1 & 0 & 1 \\ 1 & -1 & 2 \end{bmatrix} \longrightarrow \begin{bmatrix} 1 & 0 & 0 \\ 0 & 1 & 0 \\ 0 & 0 & 1 \\ 0 & 0 & 0 \end{bmatrix}$, the vectors $\mathbf{v_1} = \begin{bmatrix} 1 \\ 1 \\ 1 \\ 1 \end{bmatrix}, \mathbf{v_2} = \begin{bmatrix} 0 \\ -1 \\ 0 \\ -1 \end{bmatrix}$, and $\mathbf{v_3} =$

$\begin{bmatrix} -1 \\ 2 \\ 1 \\ 2 \end{bmatrix}$ are linearly independent, so $B = \{\mathbf{v_1}, \mathbf{v_2}, \mathbf{v_3}\}$ is a basis for $\mathbf{col}(A)$.

b. An orthogonal basis is $B_1 = \left\{ \begin{bmatrix} 1 \\ 1 \\ 1 \\ 1 \end{bmatrix}, \begin{bmatrix} 1/2 \\ -1/2 \\ 1/2 \\ -1/2 \end{bmatrix}, \begin{bmatrix} -1 \\ 0 \\ 1 \\ 0 \end{bmatrix} \right\}$.

c. An orthonormal basis is $B_2 = \left\{ \begin{bmatrix} 1/2 \\ 1/2 \\ 1/2 \\ 1/2 \end{bmatrix}, \begin{bmatrix} 1/2 \\ -1/2 \\ 1/2 \\ -1/2 \end{bmatrix}, \begin{bmatrix} -\sqrt{2}/2 \\ 0 \\ \sqrt{2}/2 \\ 0 \end{bmatrix} \right\}$.

d. $Q = \begin{bmatrix} 1/2 & 1/2 & -\sqrt{2}/2 \\ 1/2 & -1/2 & 0 \\ 1/2 & 1/2 & \sqrt{2}/2 \\ 1/2 & -1/2 & 0 \end{bmatrix}$, $R = \begin{bmatrix} 2 & -1 & 2 \\ 0 & 1 & -2 \\ 0 & 0 & \sqrt{2} \end{bmatrix}$

e. The matrix has the QR-factorization $A = QR$.

Chapter Test Chapter 6

---

1. T

2. T

3. F. $W \cap W^\perp = \{\mathbf{0}\}$

4. F. Every set of pairwise orthogonal vectors are also linearly independent.

5. F.

$$\|\mathbf{v_1}\| = \sqrt{2^2 + 1^2 + (-4)^2 + 3^2}$$
$$= \sqrt{30}$$

6. T

7. T

8. F.

$$\langle \mathbf{v_1}, \mathbf{v_2} \rangle = -4 + 1 - 8 + 3 = -8 \neq 0$$

9. F.

$$\cos \theta = \frac{\langle \mathbf{v_1}, \mathbf{v_2} \rangle}{\sqrt{30}\sqrt{10}}$$
$$= \frac{-4\sqrt{3}}{15}$$

10. F.

$$\text{proj}_{\mathbf{v_1}} \mathbf{v_2} = \frac{\langle \mathbf{v_1}, \mathbf{v_2} \rangle}{\langle \mathbf{v_1}, \mathbf{v_1} \rangle} \mathbf{v_1}$$

$$= -\frac{4}{15} \begin{bmatrix} 2 \\ 1 \\ -4 \\ 3 \end{bmatrix}$$

11. T

12. T

13. T

14. T

15. F. If $\mathbf{v_1} = \begin{bmatrix} 1 \\ 0 \\ 1 \end{bmatrix}$ and $\mathbf{v_2} = \begin{bmatrix} -1 \\ 1 \\ 1 \end{bmatrix}$, then

$$W^\perp = \mathbf{span} \left\{ \begin{bmatrix} -1 \\ -2 \\ 1 \end{bmatrix} \right\}.$$

16. T

17. T

18. T

19. F.

$$\langle 1, x^2 - 1 \rangle = \int_{-1}^{1} (x^2 - 1)dx$$

$$= -\frac{4}{3}$$

20. T

21. F.

$$\|1/2\| = \sqrt{\int_{-1}^{1} \frac{1}{4} dx} = \frac{\sqrt{2}}{2}$$

22. T

23. F. A basis for $W^\perp$ is $\{x\}$.

24. F. The only eigenvalue of the $n \times n$ identity matrix is $\lambda = 1$.

25. F. Since $\langle \mathbf{u}, \mathbf{u} \rangle = 2u_1 u_2$, the inner product can be 0 for nonzero vectors with $u_1 = 0$ or $u_2 = 0$.

26. F.

$$\langle 2\mathbf{u}, 2\mathbf{v} + 2\mathbf{w} \rangle = 2 \langle \mathbf{u}, \mathbf{v} \rangle + 4 \langle \mathbf{u}, \mathbf{w} \rangle.$$

27. T

28. T

29. T

30. F. It is the line perpendicular given by $y = -\frac{1}{2}x$.

31. T

32. T

33. F. The null space of

$$A = \begin{bmatrix} 1 & 0 \\ 2 & 1 \\ 1 & -1 \end{bmatrix} \text{ is a subset of}$$

$\mathbb{R}^2$.

34. F. Since the spanning vectors of W are linearly independent and

$$3 = \dim(\mathbb{R}^3) = \dim(W) + \dim(W^\perp),$$

then $\dim(W^\perp) = 1$.

35. T

36. F. If $\dim(W) = \dim(W^\perp)$, then the sum can not be 5.

37. T

38. T

39. F. If $\mathbf{u} = \mathbf{v}$, then $\text{proj}_{\mathbf{v}} \mathbf{u} = \text{proj}_{\mathbf{u}} \mathbf{v}$ but the vectors are linearly dependent.

40. T

Preliminaries

Exercise Set A.1

1. $A \cap B = \{-2, 2, 9\}$.

3. $A \times B = \{(a, b) \mid a \in A, b \in B\}$
There are $9 \times 9 = 81$ ordered pairs in $A \times B$.

5. $A \backslash B = \{-4, 0, 1, 3, 5, 7\}$

7. $A \cap B = [0, 3]$

9. $A \backslash B = (-11, 0)$

11. $A \backslash C = (-11, -9)$

13. $(A \cup B) \backslash C = (-11, -9)$

15.

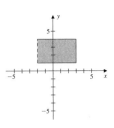

17.

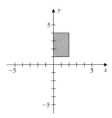

19.

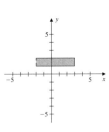

21. $(A \cap B) \cap C = \{5\} = A \cap (B \cap C)$

23.

$A \cap (B \cup C) = \{1, 2, 5, 7\} = (A \cap B) \cup (A \cap C)$

25.

$A \backslash (B \cup C) = \{3, 9, 11\} = (A \backslash B) \cap (A \backslash C)$

27. Let $x \in (A^c)^c$. Then x is in the complement of A^c, that is, $x \in A$, so $(A^c)^c \subseteq A$. If $x \in A$, then x is not in A^c, that is, $x \in (A^c)^c$, so $A \subseteq (A^c)^c$. Therefore, $A = (A^c)^c$.

29. Let $x \in A \cap B$. Then $x \in A$ and $x \in B$, so $x \in B$ and $x \in A$. Hence $x \in B \cap A$. Similarly, we can show that if $x \in B \cap A$, then $x \in A \cap B$.

31. Let $x \in (A \cap B) \cap C$. Then $(x \in A$ and $x \in B)$ and $x \in C$. So $x \in A$ and $(x \in B$ and $x \in C)$, and hence, $(A \cap B) \cap C \subseteq A \cap (B \cap C)$. Similarly, we can show that $A \cap (B \cap C) \subseteq (A \cap B) \cap C$.

33. Let $x \in A \cup (B \cap C)$. Then $x \in A$ or $x \in (B \cap C)$, so $x \in A$ or $(x \in B$ and $x \in C)$. Hence, $(x \in A$ or $x \in B)$ and $(x \in A$ or $x \in C)$. Therefore, $x \in (A \cup B) \cap (A \cup C)$, so we have that $A \cup (B \cap C) \subseteq (A \cup B) \cap (A \cup C)$. Similarly, we can show that $(A \cup B) \cap (A \cup C) \subseteq A \cup (B \cap C)$.

35. Let $x \in A \backslash B$. Then $(x \in A)$ and $x \notin B$, so $(x \in A)$ and $x \in B^c$. Hence, $A \backslash B \subseteq A \cap B^c$. Similarly, if $x \in A \cap B^c$, then $(x \in A)$ and $(x \notin B)$, so $x \in A \backslash B$. Hence $A \cap B^c \subseteq A \backslash B$.

37. Let $(x, y) \in A \times (B \cap C)$. Then $x \in A$ and $(y \in B$ and $y \in C)$. So $(x, y) \in A \times B$ and $(x, y) \in A \times C$, and hence, $(x, y) \in (A \times B) \cap (A \times C)$. Therefore, $A \times (B \cap C) \subseteq (A \times B) \cap (A \times C)$. Similarly, $(A \times B) \cap (A \times C) \subseteq A \times (B \cap C)$.

39. Let $(x, y) \in A \times (B \cap C)$. Then $x \in A$ and $(y \in B$ and $y \in C)$. So $(x, y) \in A \times B$ and $(x, y) \in A \times C$, and hence, $(x, y) \in (A \times B) \cap (A \times C)$. Therefore, $A \times (B \cap C) \subseteq (A \times B) \cap (A \times C)$. Similarly, $(A \times B) \cap (A \times C) \subseteq A \times (B \cap C)$.

Exercise Set A.2

1. Since for each first coordinate there is a unique second coordinate, then f is a function.

3. Since there is no x such that $f(x) = 14$, then the function is not onto. The range of f is the set $\{-2, -1, 3, 9, 11\}$.

5. The inverse image is the set of all numbers that are mapped to -2 by the function f, that is $f^{-1}(\{-2\}) = \{1, 4\}$.

7. Since $f(1) = -2 = f(4)$, then f is not one-to-one and hence, does not have an inverse.

9. To define a function that is one-to-one we need to use all the elements of $\mathbb{X}$ and assure that if $a \neq b$, then $f(a) \neq f(b)$. For example, $\{(1, -2), (2, -1), (3, 3), (4, 5), (5, 9), (6, 11)\}$.

11. Since

$$f(A \cup B) = f((-3, 7)) = [0, 49)$$

and

$$f(A) \cup f(B) = [0, 25] \cup [0, 49) = [0, 49),$$

the two sets are equal.

13. Since

$$f(A \cap B) = f(\{0\}) = \{0\}$$

and

$$f(A) \cap f(B) = [0, 4] \cap [0, 4] = [0, 4],$$

then $f(A \cap B) \subset f(A) \cap f(B)$, but the two sets are not equal.

15. To find the inverse let $y = ax + b$ and solve for x in terms of y. That is, $x = \frac{y-b}{a}$. The inverse function is commonly written using the same independent variable that is used for f, so $f^{-1}(x) = \frac{x-b}{a}$.

17. Several iterations give

$$f^{(1)} = f(x) = -x + c, f^{(2)}(x) = f(f(x)) = f(-x + c) = -(-x + c) + c = x,$$

$$f^{(3)}(x) = f(f^{(2)}(x)) = f(x) = -x + c, f^{(4)}(x) = f(f^{(3)}(x)) = f(-x + c) = x,$$

and so on. If n is odd, then $f^{(n)}(x) = -x + c$ and if n is even, then $f^{(n)}(x) = x$.

19. a. To show that f is one-to-one, we have that

$$e^{2x_1 - 1} = e^{2x_2 - 1} \Leftrightarrow 2x_1 - 1 = 2x_2 - 1 \Leftrightarrow x_1 = x_2.$$

b. Since the exponential function is always positive, f is not onto $\mathbb{R}$. **c.** Define $g : \mathbb{R} \to (0, \infty)$ by $g(x) = e^{2x-1}$. **d.** Let $y = e^{2x-1}$. Then $\ln y = \ln e^{2x-1}$, so that $\ln y = 2x - 1$. Solving for x gives $x = \frac{1 + \ln y}{2}$. Then $g^{-1}(x) = \frac{1}{2}(1 + \ln x)$.

21. a. To show that f is one-to-one, we have that $2n_1 = 2n_2$ if and only if $n_1 = n_2$.

b. Since every image is an even number, the range of f is a proper subset of $\mathbb{N}$, and hence, the function f is not onto. **c.** Since every natural number is mapped to an even natural number, we have that $f^{-1}(E) = \mathbb{N}$ and $f^{-1}(O) = \phi$.

23. a. Let p and q be odd numbers, so there are integers m and n such that $p = 2m + 1$ and $q = 2n + 1$. Then $f((p, q)) = f((2m + 1, 2n + 1)) = 2(2m + 1) + 2n + 1 = 2(2m + n) + 3$, which is an odd number. Hence, $f(A) = \{2k + 1 \mid k \in \mathbb{Z}\}$. **b.** $f(B) = \{2k + 1 \mid k \in \mathbb{Z}\}$ **c.** Since $f((m, n)) = 2m + n = 0 \Leftrightarrow n = -2m$, then $f^{-1}(\{0\}) = \{(m, n) \mid n = -2m\}$. **d.** $f^{-1}(E) = \{(m, n) \mid n \text{ is even}\}$ **e.** $f^{-1}(O) = \{(m, n) \mid n \text{ is odd}\}$ **f.** Since $f((1, -2)) = 0 = f((0, 0))$, then f is not one-to-one.
g. If $z \in \mathbb{Z}$, let $m = 0$ and $n = z$, so that $f(m, n) = z$.

25. Let $y \in f(A \cup B)$. Then there is some $x \in A \cup B$ such that $f(x) = y$. Since $x \in A \cup B$ with $y = f(x)$, then $(x \in A$ with $y = f(x))$ or $(x \in B$ with $y = f(x))$, so that $y \in f(A)$ or $y \in f(B)$. Hence, $f(A \cup B) \subseteq f(A) \cup f(B)$. Now let $y \in f(A) \cup f(B)$, that is, $y \in f(A)$ or $y \in f(B)$. So there exists $x_1 \in A$ or $x_2 \in B$ such that $f(x_1) = y$ and $f(x_2) = y$. Thus, there is some $x \in A \cup B$ such that $f(x) = y$. Therefore, $f(A) \cup f(B) \subseteq f(A \cup B)$.

27. Let $y \in f(f^{-1}(C))$. So there is some $x \in f^{-1}(C)$ such that $f(x) = y$, and hence, $y \in C$. Therefore $f(f^{-1}(C)) \subseteq C$.

29. Let $c \in C$. Since f is a surjection, there is some $b \in B$ such that $f(b) = c$. Since g is a surjection, there is some $a \in A$ such that $g(a) = b$. Then $(f \circ g)(a) = f(g(a)) = f(b) = c$, so that $f \circ g$ is a surjection. Next we need to show that $f \circ g$ is one-to-one. Suppose $(f \circ g)(a_1) = (f \circ g)(a_2)$, that is $f(g(a_1)) = f(g(a_2))$. Since f is one-to-one, then $g(a_1) = g(a_2)$. Now since g is one-to-one, then $a_1 = a_2$ and hence, $f \circ g$ is one-to-one.

31. Let $y \in f(A) \backslash f(B)$. Then $y \in f(A)$ and $y \notin f(B)$. So there is some $x \in A$ but which is not in B, with $y = f(x)$. Therefore $x \in A \backslash B$ with $y = f(x)$, so $f(A) \backslash f(B) \subseteq f(A \backslash B)$.

Exercise Set A.3

1. If the side is x, then by the Pythagorean Theorem the hypotenuse is given by $h^2 = x^2 + x^2 = 2x^2$, so $h = \sqrt{2}x$.

3. If the side is x, then the height is $h = \frac{\sqrt{3}}{2}x$, so the area is $A = \frac{1}{2}x\frac{\sqrt{3}}{2}x = \frac{\sqrt{3}}{4}x^2$.

5. If a divides b, there is some k such that $ak = b$ and if b divides c, there is some ℓ such that $b\ell = c$. Then $c = b\ell = (ak)\ell = (k\ell)a$, so a divides c.

7. If n is odd, there is some k such that $n = 2k+1$. Then $n^2 = (2k+1)^2 = 2(2k^2+k)+1$, so n^2 is odd.

9. If $b = a + 1$, then $(a + b)^2 = (2a + 1)^2 = 2(2a^2 + 2a) + 1$, so $(a + b)^2$ is odd.

11. Let $m = 2$ and $n = 3$. Then $m^2 + n^2 = 13$, which is not divisible by 4.

13. In a direct argument we assume that n^2 is odd. This implies $n^2 = 2k + 1$ for some integer k but taking square roots does not lead to the conclusion. So we use a contrapositive argument. Suppose n is even, so there is some k such that $n = 2k$. Then $n^2 = 4k^2$, so n^2 is even.

15. To use a contrapositive argument suppose $p = q$. Since p and q are positive, then $\sqrt{pq} = \sqrt{p^2} = p = \frac{p+q}{2}$.

17. Using the contrapositive argument we suppose $x > 0$. If $\epsilon = \frac{x}{2} > 0$, then $x > \epsilon$.

19. Contradiction: Suppose $\sqrt[3]{2} = \frac{p}{q}$ such that p and q have no common factors. Then $2q^3 = p^3$, so p^3 is even and hence p is even. This gives that q is also even, which contradicts the assumption that p and q have no common factors.

21. If $7xy \leq 3x^2 + 2y^2$, then $3x^2 - 7xy + 2y^2 = (3x - y)(x - 2y) \geq 0$. There are two cases, either both factors are greater than or equal to 0 or both are less than or equal to 0. The first case is not possible since the assumption is that $x < 2y$. Therefore, $3x \leq y$.

23. Define $f : \mathbb{R} \to \mathbb{R}$ by $f(x) = x^2$. Let $C = [-4, 4]$ and $D = [0, 4]$. Then $f^{-1}(C) = [-2, 2] = f^{-1}(D)$ but $C \nsubseteq D$.

25. If $x \in f^{-1}(C)$, then $f(x) \in C$. Since $C \subset D$, then $f(x) \in D$. Hence, $x \in f^{-1}(D)$.

27. To show that $f(A \backslash B) \subset f(A) \backslash f(B)$, let $y \in f(A \backslash B)$. This part of the proof does not require that f be one-to-one. So there is some x such that $y = f(x)$ with $x \in A$ and $x \notin B$. So $y \in f(A) \backslash f(B)$ and hence, $f(A \backslash B) \subset f(A) \backslash f(B)$. Now suppose $y \in f(A) \backslash f(B)$. So there is some $x \in A$ such that $y = f(x)$. Since f is one-to-one this is the only preimage for y, so $x \in A \backslash B$. Therefore, $f(A) \backslash f(B) \subset f(A \backslash B)$.

29. By Theorem 3, of Section A.2, $f(f^{-1}(C)) \subset C$. Let $y \in C$. Since f is onto, there is some x such that $y = f(x)$. So $x \in f^{-1}(C)$, and hence $y = f(x) \in f(f^{-1}(C))$. Therefore, $C \subset f(f^{-1}(C))$.

Exercise Set A.4

1. For the base case $n = 1$, we have that the left hand side of the summation is $1^2 = 1$ and the right hand side is $\frac{1(2)(3)}{6} = 1$, so the base case holds. For the inductive hypothesis suppose the summation formula holds for the natural number n. Next consider

$$1^2 + 2^2 + 3^2 + \cdots + n^2 + (n+1)^2 = \frac{n(n+1)(2n+1)}{6} + (n+1)^2 = \frac{n+1}{6}(2n^2 + 7n + 6)$$
$$= \frac{n+1}{6}(2n+3)(n+2) = \frac{(n+1)(n+2)(2n+3)}{6}.$$

Hence, the summation formula holds for all natural numbers n.

3. For the base case $n = 1$, we have that the left hand side of the summation is 1 and the right hand side is $\frac{1(3-1)}{2} = 1$, so the base case holds. For the inductive hypothesis suppose the summation formula holds for the natural number n. Next consider

$$1 + 4 + 7 + \cdots + (3n - 2) + (3(n+1) - 2) = \frac{n(3n-1)}{2} + (3n+1) = \frac{3n^2 + 5n + 2}{2}$$
$$= \frac{(n+1)(3n+2)}{2}.$$

Hence, the summation formula holds for all natural numbers n.

5. For the base case $n = 1$, we have that the left hand side of the summation is 2 and the right hand side is $\frac{1(4)}{2} = 2$, so the base case holds. For the inductive hypothesis suppose the summation formula holds for the natural number n. Next consider

$$2 + 5 + 8 + \cdots + (3n - 1) + (3(n+1) - 1) = \frac{1}{2}(3n^2 + 7n + 4) = \frac{(n+1)(3n+4)}{2}$$
$$= \frac{(n+1)(3(n+1)+1)}{2}.$$

Hence, the summation formula holds for all natural numbers n.

7. For the base case $n = 1$, we have that the left hand side of the summation is 3 and the right hand side is $\frac{3(2)}{2} = 3$, so the base case holds. For the inductive hypothesis suppose the summation formula holds for the natural number n. Next consider

$$3 + 6 + 9 + \cdots + 3n + 3(n+1) = \frac{1}{2}(3n^2 + 9n + 6) = \frac{3}{2}(n^2 + 3n + 2) = \frac{3(n+1)(n+2)}{2}.$$

Hence, the summation formula holds for all natural numbers n.

9. For the base case $n = 1$, we have that the left hand side of the summation is $2^1 = 2$ and the right hand side is $2^2 - 2 = 2$, so the base case holds. For the inductive hypothesis suppose the summation formula holds for the natural number n. Next consider

$$\sum_{k=1}^{n+1} 2^k = \sum_{k=1}^{n} 2^k + 2^{n+1} = 2^{n+1} - 2 + 2^{n+1} = 2^{n+2} - 2.$$

Hence, the summation formula holds for all natural numbers n.

11. The entries in the table show values of the sum for selected values of n.

n	$2 + 4 + \cdots + 2n$
1	$2 = 1(2)$
2	$6 = 2(3)$
3	$12 = 3(4)$
4	$40 = 4(5)$
5	$30 = 5(6)$

The pattern displayed by the data suggests the sum is $2 + 4 + 6 + \cdots + (2n) = n(n+1)$.

For the base case $n = 1$, we have that the left hand side of the summation is 2 and the right hand side is $1(2) = 2$, so the base case holds. For the inductive hypothesis suppose the summation formula holds for the natural number n. Next consider

$$2 + 4 + 6 + \cdots + 2n + 2(n+1) = n(n+1) + 2(n+1) = (n+1)(n+2).$$

Hence, the summation formula holds for all natural numbers n.

13. The base case $n = 5$ holds since $32 = 2^5 > 25 = 5^2$. The inductive hypothesis is $2^n > n^2$ holds for the natural number n. Consider $2^{n+1} = 2(2^n)$, so that by the inductive hypothesis $2^{n+1} = 2(2^n) > 2n^2$. But since $2n^2 - (n+1)^2 = n^2 - 2n - 1 = (n-1)^2 - 2 > 0$, for all $n \geq 5$, we have $2^{n+1} > (n+1)^2$.

15. The base case $n = 1$ holds since $1^2 + 1 = 2$, which is divisible by 2. The inductive hypothesis is $n^2 + n$ is divisible by 2. Consider $(n+1)^2 + (n+1) = n^2 + n + 2n + 2$. By the inductive hypothesis, $n^2 + n$ is divisible by 2, so since both terms on the right are divisible by 2, then $(n+1)^2 + (n+1)$ is divisible by 2. Alternatively, observe that $n^2 + n = n(n+1)$, which is the product of consecutive integers and is therefore even.

17. For the base case $n = 1$, we have that the left hand side of the summation is 1 and the right hand side is $\frac{r-1}{r-1} = 1$, so the base case holds. For the inductive hypothesis suppose the summation formula holds for the natural number n. Next consider

$$1 + r + r^2 + \cdots + r^{n-1} + r^n = \frac{r^n - 1}{r - 1} + r^n = \frac{r^n - 1 + r^n(r-1)}{r-1} = \frac{r^{n+1} - 1}{r - 1}.$$

Hence, the summation formula holds for all natural numbers n.

19. Since by Theorem 1, of Section A.1, $A \cap (B_1 \cup B_2) = (A \cap B_1) \cup (A \cap B_2)$, so the base case $n = 2$ holds. Suppose the formula holds for the natural number n. Consider

$$
\begin{aligned}
A \cap (B_1 \cup B_2 \cup \cdots \cup B_n \cup B_{n+1}) &= A \cap ((B_1 \cup B_2 \cup \cdots \cup B_n) \cup B_{n+1}) \\
&= [A \cap (B_1 \cup B_2 \cup \cdots \cup B_n)] \cup (A \cap B_{n+1}) \\
&= (A \cap B_1) \cup (A \cap B_2) \cup \cdots \cup (A \cap B_n) \cup (A \cap B_{n+1}).
\end{aligned}
$$

21.

$$
\begin{aligned}
\binom{n}{r} &= \frac{n!}{r!(n-r)!} \\
&= \frac{n!}{(n-r)!(n-(n-r))!} \\
&= \binom{n}{n-r}
\end{aligned}
$$

23. By the Binomial Theorem,

$$2^n = (1+1)^n = \sum_{k=0}^{n} \binom{n}{k}.$$

Notes

Notes

Notes

Notes

Notes